77th Annual Eastern Snow Conference 2021

Online
9 June 2021

ISBN: 978-1-7138-3925-5

Printed from e-media with permission by:

Curran Associates, Inc.
57 Morehouse Lane
Red Hook, NY 12571

Some format issues inherent in the e-media version may also appear in this print version.

Eastern Snow Conference
C/O Dr. Krystopher Chutko
117 Science Pl-Dept. Geography
Saskatoon, Sk, Canada S7N 5C8

https://www.easternsnow.org/

Additional copies of this publication are available from:

Curran Associates, Inc.
57 Morehouse Lane
Red Hook, NY 12571 USA
Phone: 845-758-0400
Fax: 845-758-2633
Email: curran@proceedings.com
Web: www.proceedings.com

Proceedings of the
77th ANNUAL

EASTERN SNOW CONFERENCE

9 June 2021

Virtual Meeting

ISSN: 0424-1932

Proceedings of the 77th Eastern Snow Conference

CONTENTS

Foreward ix

Statement of Purpose xi

Executives for the 77th Eastern Snow Conference xiii

President's Page xv

Life Members xvi

Awards xvii

Session #1: Properties of the Snowpack

Simulating Transmissivity of Thin Snow with a Photon-Tracking Radiative Transfer Model 3
TED LETCHER, JULIE PARNO, ZOE COURVILLE, JASON OLIVIER, AND LAUREN FARNSWORTH

Spatial Variation of Snow Densities over the Third Pole, Pan-Third Pole and Arctic 4
WENYU ZHAO, TINGJUN ZHANG, YIJING LIU, BENBEN LIANG, YANHUA SUN, AND QING WEI

Enabling Comprehensive Low Latency Snow Pit Data 5
PUNEETH YOGANANDA, ROGER DE ROO, AGNELO SILVA, AND RUZBEH AKBAR

Sub-Grid Scale Variability of Snow Grain Size in the ABoVE Region 6
SIDDHARTH SINGH AND ANA P. BARROS

Session #2: Snow and Ice in the Mountains

The New openAMUNDSEN Modular Snow and Hydroclimatological Modeling Framework: Application to Data from the GEWEX INARCH Rofental Catchment and ESM-SnowMIP Meteorological Stations 9
CARSTEN BECKER, FLORIAN HANZER, MICHAEL WARSCHER, AND ULRICH STRASSER

Characterizing Alpine Glacier and Lake Changes in the Cordillera Blanca, Peru from 1987 through 2020 using Multi-Sensor Remote Observations and Random Forest Classification 10
FORREST SCHOESSOW, CHANCE CARAFICE, AND ROHIT MUKHERJEE

Quantifying Precipitation Gauge Network Uncertainty in the Canadian Rockies 11
ANDRÉ BERTONCINI AND JOHN W. POMEROY

Surface Temperature and Energy Budget of Snow-Covered Complex Terrain 12
ALVARO ROBLEDANO, GHISLAIN PICARD, LAURENT ARNAUD, FANNY LARUE, AND INES OLLIVIER

Session #3: Space- and Air-Based Snow Observation

Topographical Controls on Hydrology and Microwave Behaviour of Seasonal Snowpacks: Modeling Framework and Scaling Analysis 15
YUEQIAN CAO AND ANA P. BARROS

Subtraction of Rough Soil Surface Scattering in SWE Retrieval at X and Ku Band using SnowSAR Data from SnowEx 2017 16
JIYUE ZHU, LEUNG TSANG, EDWARD KIM, AND DO-HYUK KANG

The Effects of Canopy Structure and Topography on Seasonal Changes in Surface Reflectance Pattern in the Boreal Region of Alaska – Implication for Surface Radiation Budget 17
BIBHASH NATH AND WENGE NI-MEISTER

On the Complementary Value of Space-Based Snow Observations for Snow Mass Estimation within an Observing Simulation System Experiment 18
LIZHAO WANG, BARTON A. FORMAN, SUJAY V. KUMAR, YONGHWAN KWON, PAUL GROGAN, RHAE SUNG KIM, MELISSA WRZESIEN, AND YEOSANG YOON

Session #4: Arctic Snow and Ice

Evaluating the Potential of the Snow Model Crocus driven by in situ and Recent Reanalysis Data for Arctic Applications 21
DANIELA KRAMPE, FRANK KAUKER, MARIE DUMONT, AND ANDREAS HERBER

Snow Cover Modelling over Complex Terrain of High Arctic at Point and Distributed Scales 22
HADI MOHAMMADZADEH KHANI, CHRISTOPHE KINNARD, AND ESTHER LÉVESQUE

Detailed Features of Snow Cover Structure on Hansbreen (Svalbard) in period 2008-2019 based on Radio-Echo Sounding 23
K. KACHNIARZ, M. GRABIEC, AND M. LASKA

Session #5: Measuring the Snowpack

Evaluation of LiDAR Snow Depth Estimates from Portable Consumer Devices and their Application Towards Advancing Citizen Science 27
FRASER KING AND RICHARD KELLY

Evaluating and Improving Northeastern US Snow in the National Water Model by Leveraging Advanced Mesonet Observations: Retrospective Run and Meteorological Forcing Analysis 28
PAT NAPLE, JUSTIN R. MINDER, AND THEODORE W. LETCHER

Quantifying Precipitation Undercatch in a Citizen Scientist Weather Observation Network 29
MARIA M. SILVER

Accuracy Assessment of Snow Depth Measurements in Forested and Agricultural Environments by an Unmanned Aerial Vehicle (UAV) LiDAR 30
VASANA DHARMADASA, CHRISTOPHE KINNARD, AND MICHAEL BARAËR

Discussion: Canadian Historical SWE Dataset

Canadian Historical Snow Water Equivalent Dataset (CanSWE): Recent Update (1928-2020) and Future Directions 33
VINCENT VIONNET, COLLEEN MORTIMER, MIKE BRADY, LOUISE ARMAL, AND ROSS BROWN

Poster Session

Evaluation of the Coupled Hydrology Land-Surface Model (MESH) for High-Mountain Snow and Glacier Process Simulation 37
ABBAS FAYAD AND JOHN W. POMEROY

Winter CO_2 Fluxes Measurements in Northern Environments using a Snowpack Gas Diffusion Method 38
ALEX MAVROVIC, JUHA LEMMETYINEN, CAROLINA VOIGT, JOHANN WAGNER, OLIVER SONNENTAG, AND ALEXANDRE ROY

Impact of Passive Microwave Radiometry and LiDAR Assimilation on Hydrologic Cycle Estimation 39
ALIREZA MOGHADDASI, LIZHAO WANG, BARTON A. FORMAN, AND SUJAY V. KUMAR

The Variability of Snow Density Across Ecotypes in the Low-Relief Coastal Mountains of NunatuKavut, and Nunatsiavut Labrador, Canada 40
A. FORGET, R. WAY, R. TUTTON, Y. WANG, N. LE, AND A. TRANT

Daily Forecasts of Mountain Snowpack using a Snowdrift-Permitting Model 41
CHRISTOPHER B. MARSH, VINCENT VIONNET, KEVIN R. GREEN, RAYMOND J. SPITERI, AND JOHN W. POMEROY

MODIS does not Capture the Spatial Heterogeneity of Snow Cover Induced by Solar Radiation 42
CHRISTOPHER KINNARD, HAFSA BOUAMRI, ABDELGHANI BOUDHAR, SIMON GASCOIN, LAHOUCINE HANICH, AND ABDELGHANI CHEHBOUNI

Towards the Incorporation of Adaptive Viewing in Observing System Simulation Experiments (OSSEs) to Preferentially View Snow-Covered Terrain 43
COLIN P. MCLAUGHLIN, BARTON A. FORMAN, AND LIZHAO WANG

Improving Microwave Volume Scattering Based SWE Retrieval Performance using SnowEx 2017 SnowSAR Observations 44
DO-HYUK KANG, JIYUE ZHU, EDWARD KIM, AND LEUNG TSANG

Comparison of NASA MODIS / VIIRS Cloud-Gap-Filled with other Satellite-Derived Snow-Cover Maps 45
DOROTHY K. HALL, GEORGE A RIGGS, NICOLO E. DIGIROLAMO, ANGELA M. ERB, AND CRYSTAL B. SCHAAF

Snow Satellite Mission Concept Considerations, Key Questions, and Needed Tools 52
EDWARD KIM, PAUL HOUSER, AND ANA BARROS

Scattering Mechanics of Freshwater Ice Derived Through Polarimetric Decomposition from Sledborne Scatterometers 53
G.E. GUNN, A. THOMPSON, AND J. FERGUSON

Measuring Changes in Snowpack SWE Continuously on a Landscape Scale using Lake Water Pressure 54
HAMISH D. PRITCHARD, DANIEL FARINOTTI, AND STEVEN COLWELL

History of Winter Carnival Events in College Archives and Snowfall Observations in Williamstown, MA, 1913-2010 55
HAYDEN GILLOOLY AND ALICE BRADLEY

Comparison of *in situ* Snow Depth Measurements and Impacts on Validation of Unpiloted Aerial System Lidar over a Mixed-Use Temperate Forest Landscape: A Case Study in Durham, New Hampshire, United States 56
HOLLY PROULX, ELIZABETH A. BURAKOWSKI, ADAM G. HUNSAKER, JENNIFER M. JACOBS, FRANKLIN B. SULLIVAN, MICHAEL PALACE, EUNSANG CHO, AND CAMERON WAGNER

NASA SnowEx 2020 and 2021 Campaigns in the Western U.S. 57
HP MARSHALL, CARRIE VUYOVICH, CHRIS HIEMSTRA, KELLY ELDER, MICHAEL DURAND, AND ELIAS J. DEEB

Toward Constraining Mountain Stream Flow Constituents by Combining Citizen Scientist Acquired Geochemical Tracers with Sentinel-1 SAR Time Series in Pakistan 58
JEWELL LUND, RICHARD R. FORSTER, ELIAS J. DEEB, SUMMER B. RUPPER, STEVEN J. BURIAN, YUSUF JAMEEL, HP MARSHALL, GHULAM HUSSAIN DARS, MASOOD ALI, ABDUL GHAFOOR, AND AZHAR ZAHEER

Volume Determination and Area-Volume Scaling on a Small Bolivian Cirque Glacier, Charquini SE 59
J.L. KINCAID, I.D. DOBREVA, AND A.G. KLEIN

Evaluating and Improving Northeastern US Snow in the National Water Model by Leveraging Advanced Mesonet Observations: Point Simulations and Sensitivity Experiments 60
JUSTIN R. MINDER, PAT NAPLE, AND THEODORE W. LETCHER

Exploring the History of Snow Research Through the Presentations of the Eastern Snow Conference 61
KRYSTOPHER J. CHUTKO

Microstructural Characterization of Mid-Latitude Snowpack through Micro-Computed Tomography 66
LAUREN B. FARNSWORTH AND ZOE R. COURVILLE

Using Cosmic Ray Neutrons to Estimate Snow Water Equivalent in Prairie Environments 67
MADISON WOODLEY, ERIC SPROLES, AND SAMUEL TUTTLE

Impact of the Spatiotemporal Variability of the Snowpack Conditions on Liquid Water Fluxes 68
MICHEL BARAER

Variability of Snow Depth Distributions in a Forested Mountain Basin from UAV-Lidar Remote Sensing 69
PHILLIP HARDER AND JOHN W. POMEROY

Applying the Snow Characterization with Light and Temperature (SCLT) Method to better Understand the Evolution of a Winter Snowpack 70
ROSAMOND TUTTON AND ROBERT WAY

Global Determination of Snow Cover using Remote Sensing and a Near Real Time Processing Chain 71
SEBASTIAN RÖßLER AND ANDREAS DIETZ

East *vs*. West: Contrasting Snowpack Properties in the Weddell Sea, Antarctica 72
STEFANIE ARNDY, MARCEL NICOLAUS, AND CHRISTIAN HAAS

High Resolution Spatial Estimates of Average Snow Density and Snow Water Equivalent from Differenced LiDAR Elevations and GPR Travel-Times at Grand Mesa, Colorado 73
TATE G. MEEHAN, AHMAD HOJATIMALEK, HP MARSHALL, ELIAS J. DEEB, DAN MCGRATH, RYAN WEBB, AND RANDALL BONNELL

Streamflow Generation and the Importance of Atmospheric Rivers to Annual Flooding for the Coupled Wolverine Glacier-Creek System, Kenai Mountain, Alaska 74
TODD GROTE, ALEX CRAWFORD, AND AARON JACOBS

Snow Measurements from the First Two Years of the Coastal Labrador Climate and Weather Monitoring Program 75
YIFENG WANG, ROBERT WAY, ROSAMOND TUTTON, AND JORDAN BEER

Sno-Foo Award 77

List of Attendees 79

FOREWORD

This proceedings volume contains papers presented at the 77th Eastern Snow Conference (ESC) held 9 June 2021 as a virtual meeting hosted at the University of Saskatchewan, Canada. The meeting featured oral and poster sessions on recent advances in remote sensing, approaches to snowpack measurement, snow and ice in mountainous and Arctic regions, and snowpack properties. The meeting also included an extended discussion about the Canadian Historical SWE Dataset.

The ESC is a joint United States and Canadian forum for discussing recent work on operational, applied and scientific issues related to snow and ice. It also retains an increasing interest as a symposium where novel approaches to cryospheric science of international significance are presented. The ESC has published an annual series of proceedings since 1952. Typical topics include studies of snow and ice as materials, snow removal, meteorological forecasting, river ice control, snow hydrology, snow chemistry, glaciology, remote sensing of snow and ice, and snow ecology. Membership in the ESC is open to all interested individuals and corporations. Additional copies of the current proceedings and all back issues can be obtained from the Secretary. Additional information about the Eastern Snow Conference may be found at: http://www.easternsnow.org/.

We thank members of the ESC executive committee for their continued work toward making each annual meeting a success. This includes, but is not limited to, the work of the Research Committee who evaluate all student presentations and recommend awardees for the three student prizes.

Virtual meeting support for the 2021 Eastern Snow Conference was provided by Media Productions at the University of Saskatchewan in Saskatoon, SK. Special thanks to Don Warkentin and Terry Allington for assisting with the meeting.

The 2021 meeting of the Eastern Snow Conference and these Proceedings were made possible in part by Corporate membership from Campbell Scientific Canada, Geonor, Gradient Wind Engineering, and Hoskin Scientific Ltd. Printing of the Proceedings was supported by the Centre for Hydrology, University of Saskatchewan.

GEONOR
Augusta, New Jersey, USA
http://www.geonor.com

Gradient Wind Engineering
Ottawa, Ontario, Canada
www.gradientwind.com

Centre for Hydrology
University of Saskatchewan
Saskatoon/Canmore, SK/AB
Canada
https://research-groups.usask.ca/hydrology/

The 78th Eastern Snow Conference meeting will be held 1-3 & 6-8 June 2022 in conjunction with the Canadian Meteorological and Oceanographic Society and the Canadian Geophysical Union. The meeting will be held virtually.

Krystopher Chutko and Eli Deeb
ESC Proceedings Co-Editors
University of Saskatchewan and CRREL

STATEMENT OF PURPOSE

The Eastern Snow Conference (ESC) is a joint Canadian/US organization founded in the 1940s, originally with members primarily from eastern North America. Our current members are scientists, snow surveyors, engineers, technicians, professors, students, and operational and maintenance professionals from North America, the United Kingdom, Japan, and Germany. There is a western counterpart to the ESC, the Western Snow Conference (WSC), which also is a joint Canadian/US organization.

The Eastern Snow Conference is a forum that brings the research and operations communities together to discuss recent work on scientific, applied, and operational issues related to snow and ice. The location of the conference alternates yearly between the United States and Canada, and attendees present their work by either giving a talk or presenting a poster. Most resulting papers are reviewed, edited, and published in our yearly *Proceedings of the Eastern Snow Conference*. In recent years, the ESC meetings have included sessions on snow physics, winter survival of animals, snow and ice loads on structures, river ice, remote sensing of snow and ice, and glacier processes. Volumes of the *Proceedings* can be found in libraries throughout North America and Europe, and the papers are also available through the National Technical Information Service (NTIS) in the United States and CISTI in Canada. All volumes of the Proceedings are now available at *easternsnow.org*.

*   *   *   *   *   *   *   *   *

Le Colloque sur la neige région Est (ESC) est une organisation américaine-canadienne fondée dans les années '40 et dont les membres provenaient à l'origine principalement de l'est de l'Amérique-du-Nord. Actuellement, les membres, qu'ils soient chercheurs, techniciens, ingénieurs, professeurs, étudiants, et spécialistes des services d'exploitation et d'entretien, viennent non seulement d'Amérique-du-Nord, mais aussi du Royaume-Uni, du Japon, et d'Allemagne. Le Colloque sur la neige-région Ouest (WSC), aussi une organisation américaine-canadienne, est l'homologue de l'ESC pour l'ouest nord-américain.

L'ESC est un forum qui rassemble chercheurs et responsables des services d'exploitation pour discuter des travaux récents sur les problèmes scientifiques, opérationnels, ou autres dus à la neige et à la glace. Le site de cette réunion annuelle alterne entre les États Unis et le Canada. Les participants y présentent les résultats de leurs travaux par des communications orales ou au moyen d'affiches. Ces communications, une fois revues et éditées, sont publiées dans les *Annales de l'ESC*. Dans les années récentes, les réunions de l'ESC ont inclus des sessions sur la physique de la neige, la survie hivernale de la faune, les forces exercées par la neige et la glace sur les structures et les bâtiments, la glace de rivière, la télédétection de la neige et de la glace, et les processus glaciaires. Les *Annales de l'ESC* sont accessibles dans la plupart des bibliothèques scientifiques d'Amérique-du-Nord et d'Europe. Des copies d'articles peuvent être obtenues du National Technical Information Service (NTIS) aux États Unis et son équivalent au Canada, le CISTI. Toutes les volumes des *Annales de l'ESC* sont maintenant disponibles sur *easternsnow.org*.

EXECUTIVES FOR THE 2020-21 77th EASTERN SNOW CONFERENCE

PRESIDENT
Eli Deeb
Hanover
New Hampshire, USA

PAST PRESIDENT
George Riggs
Gambrills
Maryland, USA

VICE-PRESIDENT AND PROGRAM CHAIR
Krystopher Chutko
Saskatoon
Saskatchewan, Canada

SECRETARY-TREASURER (Canada)
Krystopher Chutko
Saskatoon
Saskatchewan, Canada

SECRETARY-TREASURER (USA)
Ken Rancourt
Conway
New Hampshire, USA

ESC PROCEEDINGS EDITORS
Krystopher Chutko
Saskatoon
Saskatchewan, Canada

Eli Deeb
Hanover
New Hampshire, USA

***PHYSICAL GEOGRAPHY* EDITORS**
Mauri Pelto
Dudley
Massachusetts, USA

STEERING COMMITTEE
Craig Smith (Chair)
Saskatoon
Saskatchewan, Canada

Sean Helfrich
Suitland
Maryland, USA

Steve Howell
Toronto
Ontario, Canada

Joan Ramage
Bethlehem
Pennsylvania, USA

David Robinson
New Brunswick
New Jersey, USA

Laura Thomson
Kingston
Ontario, Canada

Carrie Vuyovich
Hanover
New Hampshire, USA

RESEARCH COMMITTEE

Barton Forman (Chair)
College Park
Maryland, USA

Laura Brown
Toronto
Ontario, Canada

James Brylawski
Augusta
New Jersey, USA

Josh King
Toronto
Ontario, Canada

Benoit Montpetit
Ottawa
Ontario, Canada

Tara Troy
Victoria
British Columbia, Canada

Sam Tuttle
South Hadley
Massachusetts, USA

WEBMASTER

Vincent Sasseville
Sherbrooke
Quebec, Canada

THE PRESIDENT'S PAGE

After the global COVID-19 pandemic led to both the postponement of the 2020 meeting and the inability to host an in-person meeting at York University (Toronto, Canada), the 77th Annual Eastern Snow Conference convened virtually on 9 June 2021. Despite not gathering in a familiar place as a community, we did enjoy the opportunity to see each other, albeit through computer displays, and engage on what brings us all together: a wide range of snow/ice science, research and applications.

For the 1-day virtual meeting, the organizing committee focused on highlighting student presentations. The scientific program included 5 oral sessions with 21 oral presentations, 30 poster submissions and a discussion session about the recent update to the Canadian historical snow water equivalent dataset (CanSWE). The discussion session included the unique history of compiling the dataset with feedback from the community about this valuable resource as well as continuing the dialogue on how to maintain it for both historical use and future updates. Although we look forward to the day when we will again meet in-person, the virtual meeting did allow many international students and attendees to participate. Despite not hosting a Banquet with a distinguished address and presentation of the Sno-foo Award, we did recognize the following students for exemplary research:

- Daniela Krampe, Alfred-Wegener-Institut, awarded the Wiesnet Medal for best student presentation entitled "Evaluating the Potential of the Snow Model Crocus driven by in situ and Recent Reanalysis Data for Arctic Applications" (Wiesnet Medal and $750 cash prize)
- Fraser King, University of Waterloo, awarded the Campbell Scientific Canada Prize for "Evaluation of LiDAR Snow Depth Estimates from Portable Consumer Devices and their Application Towards Advancing Citizen Science" ($500 cash prize)
- Madison Woodley, Syracuse University, awarded the David Miller Award for best student poster entitled "Using Cosmic Ray Neutrons to Estimate Snow Water Equivalent in Prairie Environments" ($100 cash prize)

The Executive Committee did hold a virtual meeting prior to the conference where we affirmed our commitment to Justice, Equity, Diversity, and Inclusion. We believe that it is critically important to establish these principals within our organization making them cornerstones in our decision-making process for executive, steering, and research committee nominations as well as within our nomination and selection process for awards. This is our opportunity to instill these principals in the next generation of cold region scientists and engineers.

Special thanks to Krys Chutko the Vice-President and Program Chair (for both the postponed in-person meetings and the virtual meeting!) as well as the University of Saskatchewan's Media Production for providing the virtual platform and technical support for the meeting. Thanks also to ESC Executive Committee Member Ken Rancourt who continues to provide outstanding leadership and is instrumental in the continued success of ESC meetings. As always, thank you to all of the members of the steering and research committees, notably Barton Forman, for virtually organizing the student presentation and poster judging. Likewise, our annual meetings and the ESC organization would not be successful without the participation and continued support of our corporate members: Campbell Scientific Canada, Hoskin Scientific, Geonor, and GradientWind.

As of this letter, we will likely host another virtual conference in 2022 based on uncertainty in travel across North America. Please check the ESC website for updates. Thank you for your participation and continued support. It has been a difficult few years for many of us, but we are in this together. Keep pushing the boundaries of your cryospheric science with the burgeoning methods and technologies available to us now and in the future. And share your work across broad audiences allowing us all to understand better the impacts on our changing cryosphere.

Elias Deeb
77th President, Eastern Snow Conference

LIFE MEMBERS OF THE EASTERN SNOW CONFERENCE

The Eastern Snow Conference gratefully recognizes individuals who have made lifelong contributions to the study of snow and for their support of this organization.

JAMES FOSTER

BARRY GOODISON

GERRY JONES

JOHN METCALFE

HILDA SNELLING

DONALD WIESNET

In memoriam

PETER ADAMS

DON DUNLAP

ART ESCHNER

AUSTIN LOGAN

ROBERT B. SYKES

Wiesnet Medal

The Eastern Snow Conference encourages student research through its Wiesnet Medal. This medal is presented annually to the best student paper presented at the conference.

Year	Winner	Affiliation
2021	Daniela Krampe	Alfred-Wegener-Institut
2019	Zhibang Lv	University of Saskatchewan
2018	Jiyue Zhu	University of Michigan
2017	Caroline Dolant	Université de Sherbrooke
2016	Syed Mousavi	University of Michigan
2015	Nicolas Leroux	University of Saskatchewan
2014	Justin Hartnett	Syracuse University
2013	Andreas Dietz	Earth Observation Center/DFD, Germany
2012	Elizabeth Burakowski	University of New Hampshire
2011	Kathryn Semmens	Lehigh University
2010	Simon von de Wall	University of Victoria
2009	Si Chen	Dartmouth College
2008	Chris Fuhrmann	University of North Carolina at Chapel Hill
2007	*Not awarded*	
2006	Y.C. Chung	University of Michigan
2005	M. Javan-Mashmool	Université du Québec à Chicoutimi
2004	J. Farzaneh-Dehkordi	Université du Québec à Chicoutimi
2003	Alexandre Langlois	Université de Sherbrooke
2002	Patrick Ménard	Université de Laval
2001	C. Tavakoli	Université du Québec à Chicoutimi
2000	*Not awarded*	
1999	S. Brettschnieder	Université du Québec à Chicoutimi
1998	Andrew Grundstein	University of Delaware
1997	Newell Hedstrom	University of Saskatchewan
1996	Suzanne Hartley	University of Denver
1995	Paul Wolfe	Wilfred Laurier University
1994	G.E. Mann	University of Michigan
1993	G. Devarennes	Université du Québec à Québec
1992	D.W. Cline	University of Colorado
1991	D. Samelson	Cornell University
1990	A.K. Abdel-Zaher	University of New Brunswick
1989	A. Giguere	McGill University
1988	Mauri Pelto	University of Maine
1987	Cameron Wake	Wilfred Laurier University
1986	Craig Allan	Trent University
1985	Robert Speck	Rensselaer Polytechnic Institute
1984	N.K. Kalliomaki	Laurentian University
1983	David Beresford	Trent University
1982	*Not awarded*	
1981	Jeffrey Patch	University of New Brunswick
1980	Bryan Wolfe	Trent University
1979	Margaret Leech	McGill University
1978	Michael English	Trent University
1977	Don McLaughlin & George Duggan	Rensselaer Polytechnic Institute
1976	Dwayne McMurter	Trent University
1975	Nigel Allan	Syracuse University
1974	*Not awarded*	
1973	Stan Mathewson	Trent University

David Miller Award

The David Miller Award is presented to the best student poster at the annual Conference.

Year	Winner	Affiliation
2021	Madison Woodley	Syracuse University
2019	Gaohong Yin	University of Maryland
2018	Justin Pflug	University of Washington
2017	Elizabeth Ryan	University of Maryland
2016	Oliver Wigmore	The Ohio State University
2015	Charles Papasodoro	Université de Sherbrooke
2014	Laura Thomson	University of Ottawa
2013	Nastaran Saberi	University of Waterloo
2012	Johnathan Sugg	Appalachian State University
2011	*Not awarded*	
2010	Nicolas Svacina	University of Waterloo
2009	Iliyana Dobreva	Texas A&M University
2008	Christopher Karmosky	Pennsylvania State University

Campbell Scientific Canada Prize

The Campbell Scientific Canada prize is awarded to student research showing the most innovative use of technology in the gathering of data.

Year	Winner	Affiliation
2021	Fraser King	University of Waterloo
2019	Alex Mavrovic	Université de Sherbrooke
2018	*Not awarded*	
2017	Abigail Dalton	University of Ottawa
2016	*Not awarded*	
2015	*Not awarded*	
2014	Aaron Campbell	University of Waterloo
2013	*Not awarded*	
2012	Andrew Kasurak	University of Waterloo
2011	Mathieu Beaulieu	University of British Columbia
2010	*Not awarded*	
2009	*Not awarded*	
2008	Nicholas Kinar	University of Saskatchewan
2007	Gro Lilbaek	University of Saskatchewan
2006	Gernot Koboltschnig	University of Natural Resources and Applied Life Sciences, Vienna
2005	Warren Helgason	University of Saskatchewan

Session #1: Properties of the Snowpack

Simulating Transmissivity of Thin Snow with a Photon-Tracking Radiative Transfer Model

TED LETCHER[1], JULIE PARNO[1], ZOE COURVILLE[1], JASON OLIVIER[1], AND LAUREN FARNSWORTH[1]

ABSTRACT

The interaction between visible and near-infrared radiation and individual snow particles is well described by geometric optics theory. However, this interaction is complicated enormously when considering a snowpack in aggregate due to the intricate microscopic details of the snow in ways that are not sufficiently understood to accurately model and predict radiative transfer through a snowpack in all circumstances. This has important implications for quantifying the effect of snow cover on the Earth's energy balance, determining of snow properties via remote sensing, and detecting of objects under the snow, among others. Snow has an intricate microstructure and is a highly scattering medium. Yet, most radiative transfer models rely on simplified snow grain geometries and idealized particle-size distributions to parameterize the snowpack optical properties and therefore are not able to explore in detail the impact of the snow microstructure on bulk snowpack radiative properties. In this study, we aim to understand how snow microstructure, including grain and pore space arrangement, impact the radiative transfer properties of optically thin snow with a focus on transmissivity. To accomplish this we have developed the ERDC-CRREL snow radiative transfer model (RTM) based largely on the framework developed by Kaempfer *et al.* (2007), which applies a photon-tracking model that incorporates realistic snow microstructural elements from micro-computed tomography data as input. Our model modifies this approach such that it can be expanded to snowpacks of arbitrary depth. We present an initial evaluation of this model against snow reflectance and transmittance measurements collected over painted panels inserted at various depths throughout a snowpack in Vermont during the 2020-21 winter. Microtomography samples collected coincidently with the reflectance and transmittance measurements were used to drive the radiative transfer model. Preliminary results show good agreement between modeled and observed radiative properties. The model is able to successfully reproduce known relationships between grain size and albedo.

[1] Cold Regions Research and Engineering Laboratory, 72 Lyme Rd, Hanover, New Hampshire, USA

Spatial Variation of Snow Densities over the Third Pole, Pan-Third Pole and Arctic

WENYU ZHAO[1,2], TINGJUN ZHANG[1], YIJING LIU[1], BENBEN LIANG[1], YANHUA SUN[1], AND QING WEI[1]

ABSTRACT

Snow density is essential to extensive applications for validations of satellite remote sensing products and numerical modelling outputs, surface hydrological processes, and water resource assessments. Very often, snow density was treated as a constant throughout the region and even over time due to limited ground-based measurements. The primary objective of this study is to investigate spatial changes and variations of snow density across the Third Pole, pan-Third Pole, and Arctic using *in situ* measurements of snow depth and snow water equivalent from 881 sites and stations over 70 years. The results demonstrate that snow density had a large spatial variability from the Arctic to the Third Pole. The Third Pole presented the lowest average snow density of 138 ± 61 kg/m^3, Arctic displayed the highest average snow density of 223 ± 54 kg/m^3, while average snow density for the pan-Third Pole was 159 ± 52 kg/m^3. The lowest mean snow density over the Third Pole is mainly because of frequent, fresh but non-stable snow cover throughout winter months, while the highest mean snow density across the Arctic is due mainly to stable snow cover over a whole winter with compacted and mature snow cover. On monthly basis, monthly snow density showed a significantly continuous increase from the October towards the June in Arctic. The seasonality of snowfall also changed greatly over the three areas. There was frequent light snowfall in autumn and late spring but very limited from November to February over the Third Pole, while snowfall was relatively heavier snowfall throughout the winter and spring in the Arctic. Snowfall patterns and amounts are the major difference to explain the density difference between the Third Pole and the Arctic.

[1] Key Laboratory of West China's Environment (DOE), College of Earth and Environmental Sciences, Lanzhou University, Lanzhou, 730000, China
[2] Chair in Hydrology and Climatology, Institute of Geography, Heidelberg University, Germany

Enabling Comprehensive Low Latency Snow Pit Data

PUNEETH YOGANANDA[1], ROGER DE ROO[1], AGNELO SILVA[2], AND RUZBEH AKBAR[3]

ABSTRACT

Current methods of monitoring snow water equivalent (SWE) from space, such as measuring the differential scatter darkening in the radio brightness at 19 and 37 GHz, rely on snow microphysical properties, such as grain size, in addition to the snow macroscopic properties like snow depth. The algorithms to invert the observations to SWE or snow depth are region-specific and require substantial ground truth to characterize the snow climatology of that region. And, of course, the climate is changing.

This research addresses this concern by automating the collection of snow ground truth. We will do this by integrating two technologies that already exist: 1. the SoilSCAPE system enables low-latency soil moisture data collection from areas that are large enough to be representative of a passive microwave footprint. 2. the University of Michigan snow sensor is a small, easily replicated data logging device that implants in the snowpack and logs the snow's temperature, density, and grain size, together with (yet uncalibrated) information on moisture and ambient light levels local (within a decimeter) of the sensor. The two systems can be integrated because the SoilSCAPE system uses the 900MHz ISM band for wireless communications, while the snow sensor system uses the same band for the measurement of the density and moisture of the snow. The printed circuit board for the snow sensor includes a space for the addition of a 900 MHz antenna. With the addition of the antenna to the snow sensor, the two systems' hardware are compatible. The bulk of the effort will be in implementing the software needed to allow the two systems to communicate, and thereby turn the SoilSCAPE system into one that can also monitor snow over a wide area.

Advantages of merging the two systems include giving to the snow sensor the intelligence of the SoilSCAPE system. Power management of a device embedded in the snowpack is important to preserving the snowpack properties. Judicious alterations of the measurement schedule, possible with the SoilSCAPE system, will enable rapid measurements of the snowpack when the snow is expected to be changing, and sparse sampling of the snowpack when excess power dissipation is undesired.

[1] University of Michigan, Ann Arbor, MI, USA
[2] METER Group, Inc., Pullman, WA, USA
[3] Massachusetts Institute of Technology, Cambridge, MA, USA

Sub-Grid Scale Variability of Snow Grain Size in the ABoVE Region

SIDDHARTH SINGH[1] AND ANA P. BARROS[1]

ABSTRACT

In this project we track the evolution of snow grain size and snow wetness in the top layers of the snowpack in the ABoVE domain at very high spatial resolution (e.g. 30 m) toward understanding the temporal evolution of the sub-grid scale variability and its scaling behaviour. Surface reflectance retrievals in near infrared channels from Landsat, MODIS, and VIIRS were used to estimate grain size at 30-m, 500-m, and 1-km resolution. The scaling analysis in the winter-spring transition and into the ablation season is further examined in the context of landform features, vegetation cover, and soil type. The spectral signature of different landcover features were identified and plotted simultaneously for three retrievals. The region is further subdivided in 5 study areas dominated by a particular landcover feature to carry out the study for multiple extents facilitating in bias removal. The scaling changes were studied for a particular temperature and snowfall occurrence in the region. For fresh snowfall events at high temperature the scaling is conserved across multiple support scales. However, the metamorphosized snow exhibited different scaling behaviour at high resolution. The threshold size for grain size retrieval is identified using the extreme value distribution.

[1] Duke University, Durham, NC, USA

Session #2: Snow and Ice in the Mountains

The New openAMUNDSEN Modular Snow and Hydroclimatological Modeling Framework: Application to Data from the GEWEX INARCH Rofental Catchment and ESM-SnowMIP Meteorological Stations

CARSTEN BECKER[1], FLORIAN HANZER[1], MICHAEL WARSCHER[1], AND ULRICH STRASSER[1]

ABSTRACT

openAMUNDSEN is the entirely redesigned open source version of AMUNDSEN, a physically based alpine snow model with over 20 years of development history. The new version – now available for the public – is a modular, flexible, and user-friendly framework written in Python. openAMUNDSEN is a fully distributed model, designed primarily for resolving the mass and energy balance of snow- and ice-covered surfaces in mountain regions. Typically, it is applied in areas ranging from the point to the regional scale (from hundreds up to thousands of square kilometers), using a spatial resolution of 10 - 100 m and a temporal resolution of 1 - 3 h. However, due to its modularity and flexible I/O possibilities, the potential applications are very versatile.

For the demonstration of the new model version we present publicly available data from the Rofental catchment, a glacierized GEWEX INARCH hydrological mountain research basin (1890-3770 m a.s.l.) in the Tyrolean Alps (Austria). Further, we use selected sites from ESM-SnowMIP with very different altitudes and snow hydrological conditions. Our numerical experiments show the sensitivity of the model to important parameters and its performance in the default configuration.

[1] Department of Geography, University of Innsbruck, Innsbruck, Austria

Characterizing Alpine Glacier and Lake Changes in the Cordillera Blanca, Peru from 1987 through 2020 using Multi-Sensor Remote Observations and Random Forest Classification

FORREST SCHOESSOW[1,2], CHANCE CARAFICE[1,3], AND ROHIT MUKHERJEE[2]

ABSTRACT

70% of Earth's tropical glaciers are found in the Peruvian Andes. As ice loss and lake expansion accelerate, there is a pressing need for accurate, on-demand data to inform risk management strategies due to growing geohazard risks and the depletion of water resources. However, the extent of these changes remains unquantified due to the observational limitations presented by steep terrain and dense cloud-cover. To address these challenges, we leverage four decades of NASA Earth observation data alongside recent advances in cloud computing and machine learning to characterize inter-annual changes. We introduce 1,400 total images collected by NASA's Landsat 5 and 8 satellites over the Cordillera Blanca since 1987 into Google Earth Engine where they are subjected to temporal, spatial, and cloud-filtering algorithms. 431 total images survive. Afterwards, we derive annual composite images using the median value of the high-confidence pixels preserved for each year. We calculate spectral indices for each of these annual composites that further distinguish water phases and land cover types across these glacierized alpine environments. We then use the derived indices alongside existing terrain models and Sentinel-1 synthetic aperture radar observations as feature inputs for a Random Forest (RF) classifier trained to map annual glacier and lake extent. Finally, we validate our model using a combination of high-resolution commercial satellite imagery, lidar data, and ground measures. Between 1987-2020, we observed a 27% reduction of total glacier surface area and 691% expansion of total lake surface area. Average annual ice loss measured 5.2 km^2 (down from 639.1 km^2 in 1987). This research represents two separate achievements with broad basic and applied science applicability: 1) an efficient workflow for generating accurate, cloud-free satellite imagery time-series data products on demand; and 2) the first regional assessment to quantify spatial-temporal patterns of lake evolution from 1987 to present, describe hotspots of accelerating change, and identify the emergence of potentially hazardous glacial lakes.

[1] Byrd Polar & Climate Research Center, Columbus, OH, USA
[2] Department of Geography, The Ohio State University, Columbus, OH, USA
[3] Department of Physics, The Ohio State University, Columbus, OH, USA

Quantifying Precipitation Gauge Network Uncertainty in the Canadian Rockies

ANDRÉ BERTONCINI[1] AND JOHN W. POMEROY[1]

ABSTRACT

Sparse gauging networks cause uncertainties in areal precipitation estimates, and the degree of uncertainty is especially high in mountain regions. Since snowfall is the primary input to cold regions hydrological models and is challenging to measure well, quantifying its uncertainty is crucial for hydrological estimation. The purpose of this research is to create a 30-year gauge-based rainfall and snowfall database to estimate the spatial-temporal-elevational precipitation uncertainty in the snowmelt-dominated Canadian Rockies – the headwaters of rivers that flow to three oceans. Data from 206 precipitation (rainfall+snowfall) gauges from several Canadian and US networks were utilized to create a 30-year database in the 125,601 km^2 mountain region ranging from NW Montana (MT) and S Alberta to NE British Columbia (BC). Available observed meteorological data were used for snowfall undercatch correction, and 9-km ERA-5 Land reanalysis data were utilized for infilling where observations were unavailable. The daily precipitation database was used for gauge network uncertainty estimation (standard deviation per grid-cell) using kriging aided by elevation at 2-km spatial resolution. The results show that uncertainty decreased in 87% of the region from 602 to 553 mm between 1991 and 2020. In the remaining 13%, which was near to decommissioned gauges or due to other factors, uncertainty increased from 531 to 543 mm. In 2020, uncertainty remained greatest in the remote northern part of the region (BC), and was least in the southern part of the region (MT). Uncertainty increases with elevation in most of the region, and it is greatest above 2,500 m. These findings suggest that although the uncertainty has decreased, it is still substantial in the most remote areas and highest elevations of the Canadian Rockies.

[1] Centre for Hydrology, University of Saskatchewan, Canmore, AB, Canada

Surface Temperature and Energy Budget of Snow-Covered Complex Terrain

ALVARO ROBLEDANO[1], GHISLAIN PICARD[1], LAURENT ARNAUD[1], FANNY LARUE[1], AND INES OLLIVIER[1]

ABSTRACT

The surface temperature controls the temporal evolution of the snowpack, playing a key role in physical processes such as snowmelt. It shows large spatial variations in mountainous areas because the surface energy budget is affected by the particular processes that occur in these areas, such as the modulation of the illumination by the local slope, the shadows and the re-illumination of the surface from surrounding slopes. The topography effects are often neglected in models, considering the surface as flat and smooth. Here we aim at estimating the surface temperature and the energy budget of snow-covered complex terrains in order to evaluate the role of the different processes that control the spatial variations. For this, a modelling chain is implemented to derive surface temperature from *in situ* radiometric and meteorological measurements. The main component is the rough surface ray-tracing (RSRT) model, based on a photon transport Monte Carlo algorithm to quantify the incident and reflected radiation on every facet of snow-covered areas described by a mesh. It is coupled to a surface scheme in order to estimate the energy budget. To assess the model performance, we use *in situ* measurements and satellite thermal observations (TIRS sensor aboard Landsat-8) in the Col du Lautaret area, in the French Alps. The satellite images are corrected from atmospheric effects with a single-channel algorithm. The results of the simulations show (i) an agreement between the simulated and observed surface temperature for a diurnal cycle in winter; (ii) the spatial variations of surface temperature are on the order of 5 to 10 °C between opposed slope orientations; (iii) that the importance of the few topographic effects considered here is variable, with a bias to a reference simulation of almost 1 K when being disabled. It is therefore necessary to account for the effects to estimate the spatial variations of the energy budget and surface temperature of snow-covered complex terrain.

[1] Institut des Géosciences de l'Environnement (IGE), CNRS, Univeristé Grenoble Alpes, Grenoble, France

Session #3: Space- and Air-based Snow Observations

Topographical Controls on Hydrology and Microwave Behavior of Seasonal Snowpacks: Modeling Framework and Scaling Analysis

YUEQIAN CAO[1] AND ANA P. BARROS[1]

ABSTRACT

Seasonal mountain snowpacks exhibit high spatial and temporal variability due to complex topography, heterogenous land-cover, and weather patterns. A coupled distributed snow hydrology-radiative transfer modeling framework with explicit vegetation parameters, forced by downscaled and topography-corrected reanalysis datasets and albedo at high spatio-temporal resolution (30-m, 15-min), was applied to the Senator Beck Basin (SBB), Colorado between September 1st, 2016 and July 28th, 2017. The high-resolution atmospheric forcing was evaluated against tower observation from two locations with intensive automated instruments in the SBB. Manual observations of SWE and snow depth, as well as streamflow were used to assess the seasonal march of snow accumulation and melt. The results show that the model captured the snow accumulation and melt processes over the basin well with simulated snowpack evolution and streamflow predictions consistent with the observations. Scaling analysis was subsequently applied to seasonal backscatter and brightness temperature fields at L-, C-, X-, Ku-, and Ka-bands and to snow depth and SWE toward synthesizing the signature of key topographic controls on snow hydrology and microwave behavior. Under uniform surface roughness conditions, the snowpack exhibited robust scaling behavior as measured by the time-varying spectral slope with a scale between 0.1 – 1 km independent of location within the watershed (and thus elevation) for both passive and active microwave emissions at all frequencies. To separate contributions of volumetric scattering from surface one at different frequencies and different seasons, a large sensitivity experiment was conducted using synthetically generated surface roughness patterns. A joint analysis of passive and active microwave hysteresis vis-à-vis fractional volume scatter provides guidance for SWE retrieval.

[1] Duke University, Durham, NC, USA

Subtraction of Rough Soil Surface Scattering in SWE Retrieval at X and Ku Band using SnowSAR Data from SnowEx 2017

JIYUE ZHU[1], LEUNG TSANG[1], EDWARD KIM[2], AND DO-HYUK KANG[3]

ABSTRACT

The physical basis of snow water equivalent (SWE) retrieval using X and Ku band radar observations is that the volume scattering of snow is assumed to increase with SWE. However, for small to moderate SWE, the total radar backscattering signal at X and Ku band contains effects of rough surface scattering. Therefore, rough surface scattering degrades sensitivity to SWE. To enhance the sensitivity and increase the correlation of X and Ku band radar observations with SWE, we propose to subtract out the rough surface effect. The subtraction procedure is as follows. Lower frequency radar backscattering data at L and C band are used to retrieve rms heights h and soil permittivity. The retrievals are based on lookup tables of rough surface scattering at L and C band. Then retrieved rms heights and permittivity are used to calculate surface scattering at X and Ku band. The volume scattering components of snow are obtained by subtracting out surface scattering from radar observations. To apply this procedure, we investigated using rough surface scattering results of physical models from L to Ku band. In this paper, we present results of soil surface scattering from L to Ku band based on the numerical solutions of Maxwell's equations in 3D (NMM3D) full wave simulations. Simulations results are plotted as a function of frequency for fixed rms heights and soil permittivity. In the past, physical model results have been limited within the range of $kh < 3$, where k is the free space wavenumber of the microwave signals. Full wave simulations have been limited in the past to $kh < 1.5$. In plains, rms height is up to 3 cm. In mountainous areas, rms height is up to 5cm. At a Ku frequency of 17.2 GHz, rms height of 5cm means $kh = 18$. Therefore, the past full wave simulations with $kh < 1.5$ are too limited. In this paper, we present results for kh up to 15. We simulate results for a wider range of rms heights from L to Ku band. We also perform and validate the subtraction procedure with L band UAVSAR and X and Ku band SnowSAR data from the SnowEx 2017 campaign.

[1] Radiation Laboratory, Department of Electrical Engineering and Computer Science, The University of Michigan, Ann Arbor, MI, USA.
[2] NASA Goddard Space Flight Center, Greenbelt, MD, USA.
[3] ESSIC, University of Maryland, College Park, MD, USA

The Effects of Canopy Structure and Topography on Seasonal Changes in Surface Reflectance Pattern in the Boreal Region of Alaska – Implication for Surface Radiation Budget

BIBHASH NATH[1] AND WENGE NI-MEISTER[1]

ABSTRACT

Forests play a critical role in regulating the world's climate. They sequester atmospheric carbon and maintain a nice energy balance of incoming and outgoing solar radiation. The variability in forest canopy characteristics create uncertainty in the estimation and modelling of Earth's radiation budget and understanding surface radiative processes within canopies. Accurate knowledge of energy balance helps better quantification of surface radiation budget and modelling prediction of climate. We studied the small-scale variability (at 30-m resolution) in land cover classes, canopy characteristics, and terrain conditions and its relationship to seasonal changes in surface reflectance pattern in the boreal regions of Alaska. We combined Landsat 8 surface reflectance and LiDAR-derived canopy matrices to examine the existence of any relationships.

The canopy characteristics and canopy structural complexity (rugosity) varies with land cover classes. Tree heights and rugosity are positively correlated. However, tree heights and rugosity are higher in deciduous forests in comparison with evergreen forests. Our exploratory data analysis suggests that the presence of snow during winter months increases visible and near-infrared reflectance in forested areas, however no significant variation in visible and near-infrared reflectance was observed during summer months (May and August). The results further suggest that the surface reflectance from visible, near-infrared, and shortwave-infrared wavelength changes with seasons, topography, canopy characteristics, and land cover classes. Large variability and decreasing visible and near-infrared reflectance in relation to increasing canopy cover, tree heights and rugosity during March could be associated with the creation of high surface heterogeneity due to snow accumulation, especially in the region dominated by evergreen forests. Higher rugosity creates surface heterogeneity because of uneven tree heights and consequently greater absorption of lights due to canopy shades.

The detailed understanding of canopy characteristics and terrain conditions and its relationship to surface reflectance patterns at a small spatial scale could help parametrization of climate change models.

[1] Department of Geography, Hunter College of the City University of New York, NY, USA

On the Complementary Value of Space-Based Snow Observations for Snow Mass Estimation within an Observing Simulation System Experiment

LIZHAO WANG[1], BARTON A. FORMAN[1], SUJAY V. KUMAR[2], YONGHWAN KWON[2,3], PAUL GROGAN[4], RHAE SUNG KIM[2,5], MELISSA L. WRZESIEN[2,5], AND YEOSANG YOON[2,6]

ABSTRACT

Snow plays a key role in the freshwater supply of many watersheds located at high elevations or high latitudes. However, despite this importance, we do not know exactly how much snow exists in the world. One challenge in quantifying snow mass using space-based sensors is that no single sensor can accurately measure all types of snow during all times of the year. That is, each type of sensor has its own strengths (and limitations) that must be considered as part of a suite of sensors in order to accurately measure global snow mass.

In this study, we develop an observing simulation system experiment (OSSE) in order to explore the individual contribution of a complementary suite of space-borne sensors. In the experiment, we first generate a "synthetic truth" of SWE and snow depth using the NoahMP-4.0.1 land surface model within the NASA Land Information System (LIS). Afterwards, synthetic SWE and snow depth retrievals from one PMW radiometer and one LiDAR, respectively, are generated by applying a space-time subsampler that provides an accurate representation of what each sensor can "see" at any point of time on Earth. Next, a realistic amount of observation error as a function of dense/sparse forest, deep/shallow snow, and dry/wet snow is then added to the synthetic retrievals. The synthetic retrievals are then assimilated into NoahMP-4.0.1 in order to help quantify the added value that each type of snow retrieval has on the land surface model performance. The results from this OSSE help reveal the complementary nature of PMW radiometer and LiDAR sensors, and in turn, can help aid mission planners in determining how to get the most observational "bang for the buck" in the selection of a future snow mission.

[1] Civil and Environmental Engineering, University of Maryland, College Park, MD, USA
[2] Hydrological Sciences Laboratory, NASA Goddard Space Flight Center, MD, USA
[3] Earth System Science Interdisciplinary Center, University of Maryland, College Park, MD, USA
[4] Stevens Institute of Technology, School of Systems and Enterprises, Hoboken, NJ, USA
[5] Universities Space Research Association (USRA), Columbia, MD, USA
[6] Science Applications International Corporation, McLean, VA, USA

Session #4: Arctic Snow and Ice

Evaluating the Potential of the Snow Model Crocus driven by *in situ* and Recent Reanalysis Data for Arctic Applications

DANIELA KRAMPE[1], FRANK KAUKER[1,2], MARIE DUMONT[3], AND ANDREAS HERBER[1]

ABSTRACT

In the Arctic, snow covers the ground for most of the year and affects flora, fauna, and local conditions but also has global impacts e.g. due to the snow-albedo effect and the impact from snow on near-surface atmospheric conditions. Previous studies show that snow can alter circulation patterns and precipitation regimes. Despite its importance, reliable and detailed *in situ* data of snow depth and physical snow properties in the Arctic are limited in time and space. The same is true for meteorological *in situ* data. These are needed to drive snow models to compensate for missing snow measurements.

To overcome these limitations, we first evaluate how well the recent ECMWF atmospheric reanalysis ERA5 represents atmospheric *in situ* measurements at Villum Research Station in northeast Greenland. Second, we use ERA5 and the atmospheric *in situ* measurements to force the alpine snow model Crocus and investigate the model performance for an Arctic site.

As atmospheric conditions in the Arctic and the European Alps are different (e.g. colder air temperatures, stronger winds), leading to differences in the snow stratigraphy, we introduce a new parametrization for the density of freshly fallen snow.

Comparison of atmospheric variables from ERA5 with *in situ* measurements results in good agreement except for biases in precipitation, wind speed, and direction. We attribute these biases to the resolution of ERA5, which is too coarse to resolve local topography adequately. However, simulations forces with ERA5 represent measured snow depth deter than simulations forces with atmospheric *in situ* measurements. The performance of Crocus to simulate snow depths is satisfying, but there are discrepancies between measured and simulated snow stratigraphy. Using the newly introduced fresh-fallen snow density parametrization leads to improvements in the simulations, but deviations remain. Nevertheless, both the ERA5 reanalysis and Crocus show high potential for future snow modeling in the Arctic and are great tools to get a more comprehensive picture of atmospheric and snow conditions in the strongly under-observed Arctic.

[1] Alfred-Wegener-Institut Helmholtz-Zentrum für Polar-und Meeresforschung, Bremerhaven, Germany
[2] Ocean Atmosphere Systems GmbH, Hamburg, Germany
[3] Université Grenoble Alpes, Université de Toulouse, Météo-France, CNRS, CNRM, Centre d'Études de la Neige, Grenoble, France

Snow Cover Modelling over Complex Terrain of High Arctic at Point and Distributed Scales

HADI MOHAMMADZADEH KHANI[1,2,3], CHRISTOPHE KINNARD[1,2,3], AND ESTHER LÉVESQUE[1,2,3]

ABSTRACT

Snow is a critically important and rapidly changing feature of the High Arctic that is significantly influenced by wind redistribution. However, the shortage of information on snow redistribution at fine spatial scales across high latitudes imposes a major limitation on permafrost and climate model projections under climate change scenarios. In this research, we examine the capacity of the physically-based GEOtop model to simulate snow dynamics at the point and basin scales in a tundra landscape on Bylot Island in the Canadian High Arctic. The simulated snow cover was compared with snow depth measured at an automated weather station since 2010. The model reproduced the physical features of snow cover fairly and showed a significant consensus with the recorded data. The result of this study showed significant spatial variability of snow cover due to blowing snow and losses from blowing snow sublimation. Snow cover properties (depth and SWE) were mostly affected by key parameters linked with albedo and by the accuracy of the input precipitation.

[1] Université du Quebec à Trois-Rivières, Trois-Rivières, QC, Canada

[2] Centre for North Studies (CEN), Québec City, QC, Canada

[3] Research Centre for Watershed-Aquatic Ecosystem Interactions (RIVE), Université du Quebec à Trois-Rivières, Trois-Rivières, QC, Canada

Detailed Features of Snow Cover Structure on Hansbreen (Svalbard) in period 2008-2019 based on Radio-Echo Sounding

K. KACHNIARZ[1], M. GRABIEC[1], AND M. LASKA[1]

ABSTRACT

Snow cover monitoring on Hansbreen (SW Svalbard) by ground-penetrating radar (GPR) is one of the longest and the most comprehensive on Svalbard. Snow depth is frequently the only feature derived from high-frequency radio-echo sounding. In this study, also physical characteristics of snow cover structure have been analyzed. We demonstrate that based on GPR data and a single snow pit, it is possible to extrapolate the detailed characteristics of snow cover to the entire accumulation zone. Layers determined on the GPR profiles are validated by assigning them attributes from snow pit analysis.

Field studies were carried out on at the end of the accumulation seasons 2008 – 2019. The snow cover stratification and their spatiotemporal changes show the relations to the parameters such as shape, size, hardness, and density. The predominant type of snow in specific seasons and their cause have been shown, and the snow water equivalent (SWE) have been calculated. Obtained results were combined with meteorological data from an automatic weather station installed on Hansbeen and by the Polish Polar Station in Hornsund. It enabled referring to temperature gradients, sequences of thaws, refreezing, and rain-on-snow episodes that favour forming the melt-freeze crusts within snow layers. The influence of hard layers on snow metamorphism and the density/SWE has been discussed. The method used in this study may improve the calculation of winter mass balance and contribute to hydrological models.

[1] University of Silesia, Centre for Polar Studies, Katowice, Poland

Session #5: Measuring the Snowpack

Evaluation of LiDAR Snow Depth Estimates from Portable Consumer Devices and their Application Towards Advancing Citizen Science

FRASER KING[1] AND RICHARD KELLY[1]

ABSTRACT

Snow is a critical contributor to the regional water-energy budget with impacts to springtime flooding and water resource management practices. *In situ* snow depth measurements are incredibly useful reference observations as inputs to hydrologic models and reanalysis. However, Canadian measurement station counts have declined by over 70% since 1990, and the current observational network is sparsely distributed with large unobserved gaps. Laser altimetry (LiDAR – Light Detection and Ranging) is a remote sensing technique that has demonstrated promising results at mapping changes in snow depth, but the expense of purchasing and transporting traditional LiDAR equipment has generally limited their use to large institutions. We demonstrate that the combination of vertical cavity surface-emitting lasers (VCSELs) and single-photon avalanche diodes (SPADs) installed on the iPhone 12 Pro comprise a portable LiDAR sensor that acts as a real-time, handheld measurement instrument for accurately observing changes in snow depth. Two independent field experiments in southern Ontario found that the iPhone LiDAR measurements were able to accurately capture daily changes in snow depth when compared to *in situ* snow ruler measurements. *In situ* and LiDAR comparisons of $n = 50$ days at measurements site 1 exhibit a correlation of $r \approx 0.98$, mean absolute bias less than 1 mm, and an RMSE of approximately 5 mm. Similar positive agreement was also noted at the second field study site for $n = 10$ measurements over the same period. As LiDAR sensors become ubiquitous components of future smartphones, their capabilities as portable snow depth measurement instruments cannot be understated. The high accuracy of the LiDAR sensor suggests that a mobile application could be developed which allows users to quickly scan a snow-covered area before and after a snowfall event and consequently use this data to aid in filling current observational gaps through a citizen science-based approach of measuring local-scale changes in snow depth.

[1] Department of Geography & Environmental Management, University of Waterloo, Waterloo, ON, Canada

Evaluating and Improving Northeastern US Snow in the National Water Model by Leveraging Advanced Mesonet Observations: Retrospective Run and Meteorological Forcing Analysis

PAT NAPLE[1], JUSTIN R. MINDER[1], AND THEODORE W. LETCHER[2]

ABSTRACT

The NOAA National Water Model (NWM) is used for operational forecast of hydrology, including surface snowpack, across the continental United States (US). It initializes surface snowpack conditions using a simple deterministic analysis cycling every hour and represents snowpack physics using the Noah-MP land surface model. However, performance of the NWM at simulating snow over the northeastern United States is poorly characterized, in part due to limitations in the observational network. Our group is working to improve snow state initialization and prediction within the NWM by leveraging observations from the advanced New York State Mesonet (NYSM; *http://nysmesonet.org/*), which includes 126 stations with detailed meteorology measurements and sub-networks with additional snowpack water equivalent (SWE) and surface energy flux measurements.

As part of this larger effort, we are quantifying biases in NWM-simulated snow and evaluating the link between snow and meteorological forcing biases. This presentation evaluates two versions (v2.0 and v2.1) of NWM retrospective runs, forced by meteorological analyses, against NYSM station observations and New York Snow Survey manual snow course data (*https://www.nrcc.cornell.edu/regional/snowsurvey/snowsurvey.html*). Simulated snow depth for the 2017-2018 snow season is evaluated against NYSM measurements. Simulated SWE from 2010-2018 is evaluated against New Your Snow Survey manual snow course measurements to validate biases found in simulated snow depth. Precipitation, temperature, and incoming shortwave radiation from forcing datasets are evaluated against NYSM observations to determine the effects of meteorological forcing biases on snow biases.

NWM v2.0 shows substantial negative snow depth biases and v2.1 shows moderately negative snow depth biases compared to NYSM observed values. Modeled SWE shows similar biases when compared with New York Snow Survey measurements. Meteorological forcing shows negative precipitation, positive temperature, and positive incoming shortwave radiation biases with larger magnitude biases seen in the forcing for NWM v2.0. These forcing biases decrease modeled snow depth by reducing accumulation and increasing ablation. Snow depth biases and precipitation forcing biases show statistically significant positive correlations, especially during accumulation events. Our results indicate that, despite being forced by analyses of meteorological observations, meteorological forcing biases contribute significantly to biases in NWM retrospective runs, complicating efforts to isolate biases in model physics.

---

[1] University at Albany, State University of New York, Albany, NY, USA
[2] Cold Regions Research and Engineering Laboratory, U.S. Army Engineer Research and Development Center, Hanover, NH, USA

Quantifying Precipitation Undercatch in a Citizen Scientist Weather Observation Network

MARIA M. SILVER[1]

ABSTRACT

Precipitation undercatch is a phenomenon that occurs when the precipitation measurement taken from the gauge is less than the true amount of precipitation. Many studies have analyzed this, but few have estimated undercatch values for 4" rain gauges. The goal of this project was to analyze what factors contribute to significantly greater undercatch and to derive adjustment formulas that estimate gauge undercatch in 4" rain gauges. Using data collected from the Community Collaborative Rain, Hail, and Snow Network (CoCoRaHS) database, I compared the daily snow water equivalent (SWE) measurements taken from a 4" gauge to daily SWE measurements taken from a snowboard – a flat, white surface placed on the ground. Snowfall catch efficiencies can vary greatly from station to station, ranging from around 25% to above 90%. A catch efficiency means that for every inch of precipitation, a rain gauge will measure that percent of an inch of precipitation. After dividing the measurements into categories based on variables that could affect undercatch (wind, season, temperature, year), I concluded that wind and temperature significantly decrease catch efficiency. Adjustment formulas were derived to quantify undercatch. These results and formulas provide empirical data that can quantify gauge undercatch.

[1] Broomfield High School, Broomfield, CO, USA

Accuracy Assessment of Snow Depth Measurements in Forested and Agricultural Environments by an Unmanned Aerial Vehicle (UAV) LiDAR

VASANA DHARMADASA[1], CHRISTOPHE KINNARD[1], AND MICHEL BARAËR[2]

ABSTRACT

With recent advances in remote sensing techniques, UAV LiDAR has been increasingly used to monitor snow depths. Due to its improved accuracy and high penetration ability through canopy gaps, LiDAR techniques have proven to be effective in capturing snow depth even in forested environments. However, it is not without errors. In general, careful system calibration and post-processing can reduce some navigational errors and errors posed by the system. In snow depth mapping or any survey-grade mapping, it is important to assess if the remaining error is within the accepted sensor accuracy and/or if further refinements (e.g. strip aligning) are required. This study demonstrates the procedure implemented to assess the absolute and relative accuracy of LiDAR data (system calibrated and post-processed) in two paired deciduous forested and agricultural sites and a boreal forested site in southern Quebec. We used circular-shaped elevated (1 m diameter) and square-shaped flat (0.5 x 0.5 m^2) ground control points (GCPs) in summer and winter surveys to assess the absolute accuracy of the LiDAR point cloud. Relative accuracy was determined by a repeat flight over one surveying block. Estimated absolute and relative errors were within the accepted accuracy of the LiDAR (~5 cm and ~7 cm, respectively). Validation of LiDAR-derived snow depths with ground-based measurements was also in good agreement. We also tested BayesStripAlign 2.17 software to correct the misalignment between overlapping flight strips over vertical (z) direction. However, the results suggest that the significant improvement of relative accuracy between overlapping flight strips made by this technique was at the cost of the absolute accuracy of LiDAR data. This phenomenon is also confirmed by the degraded performance of strip-aligned snow depths with ground-based measurements. In conclusion, in this study, the system calibrated, post-processed point cloud is adequate to provide reasonable centimeter-level accurate snow depth maps.

[1] Université du Quebec à Trois-Rivières, Trois-Rivières, QC, Canada
[2] Hydrology, Climate and Climate Change (HC[3]) Laboratory, École de technologie supérieure, University of Quebec, Canada

Discussion: Canadian Historical SWE Dataset

Canadian Historical Snow Water Equivalent Dataset (CanSWE): Recent Update (1928 – 2020) and Future Directions

VINCENT VIONNET[1], COLLEEN MORTIMER[2], MIKE BRADY[2], LOUISE ARMAL[3], AND ROSS BROWN[2]

ABSTRACT

Reliable and timely *in situ* information of snow water equivalent (SWE) – the depth of water that would be produced if all the snow melted – is critical for water management, flood forecasting, climate monitoring, evaluation of hydrological and land surface models, and for applied research. The Canadian historical SWE dataset (CanSWE) combines manual and automated pan-Canadian SWE observations collected by national, provincial, and territorial agencies as well as hydropower companies. Snow depth and derived bulk snow density are also included when available. This update supersedes the previous Canadian Historical Snow Survey dataset published by Brown *et al.* (2019), and this presentation describes the efforts made to correct metadata, remove duplicate observations, and quality control records. The updated dataset, compiled from 15 different sources, includes SWE information for all provinces and territories that measure SWE. Data were updated to July 2020 and new historical data from the Government of Northwest Territories, Government of Newfoundland and Labrador, Saskatchewan Water Security Agency, and Hydro Quebec were included. CanSWE includes over one million SWE measurements from 2607 different locations across Canada over the period 1928 – 2020. It is publicly available on the Open Canada website.

This presentation will be followed by a discussion with the snow community to identify additional snow datasets that could be included in CanSWE and to clarify the community need for future regular updates of CanSWE.

[1] Meteorological Research Division, Environment and Climate Change Canada, Dorval, QC, Canada
[2] Climate Research Division, Environment and Climate Change Canada, Toronto, ON, Canada
[3] Coldwater Laboratory, University of Saskatchewan, Canmore, AB, Canada

Poster Session

Evaluation of the Coupled Hydrology Land-Surface Model (MESH) for High-Mountain Snow and Glacier Process Simulation

ABBAS FAYAD[1] AND JOHN W. POMEROY[1]

ABSTRACT

The objectives of this study are (1) to examine the performance of the snow and glacier routines in the coupled hydrology land-surface model (MESH), that incorporates the Canadian Land Surface Scheme (CLASS); and (2) to quantify the influence of physical parameterization and meteorological forcing on snowpack and glacier simulations in high mountains. Of particular interest in MESH is its ability to simulate blowing snow, snow interception by canopy, and glacier ice energetics. Here, a physically-based parameterization approach was used with parameters selected based on the understanding of the physical system with no calibration. The model was configured in single column mode and forced with multiple years (163 years in total) of hourly *in situ* meteorological observations and run at eight alpine research sites in the Canadian Rockies that are part of the Canadian Rockies Hydrological Observatory (CRHO) and five alpine sites from the Earth System Model-Snow Model Intercomparison Project (ESM-SnowMIP). The ensemble of these sites represent alpine ridges, alpine forests and clearings, alpine tundra, grassland meadows, a barren moraine, and a glacier. Observations of shortwave and longwave irradiance were adjusted for elevation to the single column model site. A process-based approach was used to evaluate how realistically the model simulates snowpack evolution and changes in snow and ice energetics. The model represented snow accumulation, snowmelt onset, and ablation rates quite well at all sites. The modelled snow interception and sub-canopy snow routines were able to accurately simulate snowpack under forest canopy. At the wind-blown alpine ridges, an adequate representation of the snowpack evolution was possible when using MESH blowing snow algorithms. The glacier routines of the model were able to simulate ice ablation on a glacier reasonably well. Falsification of the forcing meteorology by not accounting for terrain impacts on radiative forcing and for elevational influences on precipitation and temperature resulted in large difference in simulations of snowpack evolution and melt contribution from glacier ice and so show the importance of including slope, aspect, and elevation in hydrological land surface schemes applied for mountain snow and ice simulation.

[1] Centre for Hydrology, University of Saskatchewan, Canmore, AB, Canada

Winter CO_2 Fluxes Measurements in Northern Environments using a Snowpack Gas Diffusion Method

ALEX MAVROVIC[1,2,3], JUHA LEMMETYINEN[4], CAROLINA VOIGT[1,5], JOHANN WAGNER[3], OLIVER SONNENTAG[1,5], AND ALEXANDRE ROY[1,2]

ABSTRACT

In arctic and boreal regions, the amount of carbon dioxide (CO_2) released from the soil outside of the growing season is highly uncertain. The insulating properties of snow allow the ground to maintain soil temperatures high enough to allow soil respiration, thus generating CO_2 emissions from the soil to the atmosphere despite the freezing air temperatures. Although winter soil respiration is weak, winter covers most of the year in northern environments. As a result, the winter soil respiration contribution to the annual carbon balance for these large environments cannot be neglected in order to determine whether these regions are net carbon sources or sinks. The harsh environmental conditions combined with the challenge to access the snow-covered ground surface without altering the snowpack present several technical difficulties for the measurement of winter soil respiration. Here we present preliminary results obtained in arctic and boreal environments using a snowpack gas diffusion method. This technique consists of extracting air samples within the snowpack with syringes in order to measure CO_2 concentration gradient to estimate CO_2 diffusion through the snowpack and, thus, the CO_2 fluxes. Snowpack stratigraphy and microstructure were also retrieved to quantify the impact of snow characteristics on winter CO_2 emission. The method makes it possible to estimate the CO_2 fluxes between the soil and the atmosphere without altering the snowpack. It is also an efficient technique to study the spatial variability of CO_2 fluxes. The presentation will show the spatial variability of winter CO_2 emission along snow characteristics across two boreal forest sites (Forêt Montmorency and Havipak Creek) and two tundra sites (Trail Valley Creek and Cambridge Bay).

[1] Centre d'études Nordiques, Quebec, Canada
[2] Université du Québec à Trois-Rivières, Trois-Rivières, QC, Canada
[3] Polar Knowledge Canada, Cambridge Bay, NU, Canada
[4] Finnish Meteorological Institute, Helsinki, Finland
[5] Université de Montréal, Montréal, QC, Canada

Impact of Passive Microwave Radiometry and LiDAR Assimilation on Hydrologic Cycle Estimation

ALIREZA MOGHADDASI[1], LIZHAO WANG[1], BARTON A. FORMAN[1], AND SUJAY V. KUMAR[2]

ABSTRACT

Enabling accurate estimation of hydrologic states and fluxes enhances the ability to monitor changes in the terrestrial hydrologic cycle and freshwater storage in general. Although land surface models aim to provide details on land atmosphere interactions, the lack of physical representation in these models limits their accuracy. Assimilation of satellite observations, on the other hand, is one way to reduce the errors and uncertainties in those model estimates.

This study investigates the added value associated with assimilating snow observations from passive microwave (PMW) and LiDAR sensors into the NoahMP-4.0.1 land surface model and their subsequent cascading effect on portions of the hydrologic cycle other than snow such as soil moisture and vegetation. In order to do this, a realistic representation of the satellite measurements (a.k.a., synthetic observations) for a given type of sensor is first derived by applying a space time subsampler that accounts for differences in sensor swath width, platform inclination angle, and platform altitude. These synthetic observations – with an emphasis on terrestrial snow – obtained from a variety of different sensor types and different sensor error characteristics are then assimilated into the NoahMP-4.0.1 model using an Ensemble Kalman Filters. Results from these data assimilation (DA) experiments reveal the extent of improvement in terms of the estimation of various hydrologic states and fluxes such as soil moisture, vegetation growth, surface runoff, and terrestrial water storage relative to model output when such observations are not assimilated. Results presented will quantify how observations related to terrestrial snow eventually cascade down into other components of the hydrologic cycle thereby providing benefits to other hydrologic states and fluxes of interest to hydrologists and water resource managers.

[1] Civil and Environmental Engineering, University of Maryland, College Park, MD, USA
[2] Hydrological Sciences Laboratory, NASA Goddard Space Flight Center, MD, USA

The Variability of Snow Density Across Ecotypes in the Low-Relief Coastal Mountains of NunatuKavut, and Nunatsiavut Labrador, Canada

A. FORGET[1], R. WAY[1], R. TUTTON[1], Y. WANG[1], N. LE[2], AND A. TRANT[2]

ABSTRACT

Decades of northern science has shown that the ground thermal regime of discontinuous permafrost is significantly influenced by the distribution and characteristics of seasonal snow cover. Although coarse snow metrics such as area and duration can be derived from satellite imagery, understanding the thermal impacts of snow on ground temperatures requires collecting field-intensive information, including snow thickness and density. Scale-related challenges due to local snow cover variability would be most significant in high snowfall regions with extensive wind redistribution, such as the low-relief coastal mountains in Labrador. This study presents early results from snow monitoring undertaken across forest-tundra transitions at two research basins (NBH & PRH) located near Nain, Nunatsiavut and Pinware, NunatuKavut. At both sites, winter and summer field data collection was undertaken in 2019 (July) and 2020 (March), with the latter field season including randomly distributed measurements of snow depth and snow density across different ecotypes. Remotely Piloted Aircraft (RPA) image acquisitions were also completed in March 2020 at both sites and in July 2019 at PRH, allowing for the generation of high-resolution digital surface models and derived metrics (vegetation height, snow depth) at PRH. An RPA-derived machine learning-based ecotype classification was used in conjunction with RPA estimated snow depths and a simple snow density model to spatially represent snow water equivalent (SWE) measurements across the entire PRH research basin. Our preliminary results show that the distribution and density of snow cover at the research basins is strongly linked to topographic position and ecotype, highlighting the importance of prevailing winds in coastal mountains. Our analysis also showed that tall vegetation was associated with the highest overall SWE values but the largest range in SWE (0 to 23 cm) thus reflecting large within ecotype heterogeneity Outputs from this projects will inform ground thermal and hydrological modelling throughout coastal Labrador.

[1] Northern Environmental Geoscience Laboratory, Department of Geography, Queen's University, Kingston, ON, Canada
[2] Trant Ecological Legacy Lab, School of Environment, University of Waterloo, Waterloo, ON, Canada

Daily Forecasts of Mountain Snowpack using a Snowdrift-Permitting Model

CHRISTOPHER B. MARSH[1,2], VINCENT VIONNET[3], KEVIN R. GREEN[4], RAYMOND J. SPITERI[1,4], AND JOHN W. POMEROY[1,2]

ABSTRACT

Ongoing work to simulate the substantial spatial and temporal variability in energy and mass associated with cold regions hydrology has resulted in the Canadian Hydrological Model (CHM), a distributed model development framework. It is capable of a multi-scale basin discretization to permit greater spatial resolution where it is warranted. CHM includes most snow wind redistribution, interception, sublimation, accumulation, and ablation processes as well as sophisticated wind and energetics calculations for complex terrain and forest canopies.

Recent work evaluated the capacities of the above-treeline snow redistribution routines in CHM against lidar snow depth observations for a ~1000 km^2 mountain domain in the Canadian Rockies. These simulations were conducted at a 'snowdrift-permitting' scale (200 m or finer) using a pre-processing step to pre-compute wind fields. This approach allowed for efficient downscaling of the meteorological forecasts of the High-Resolution Deterministic Prediction System (HRDPS) forecasts from Environment and Climate Change Canada.

Large-extent estimates of snow mass are of significant value for water resource estimates in complex mountain terrain. An example of implementing CHM for daily forecasting purposes as the core of the Snowcast snow prediction system in the Canadian Rockies is shown. Specifically, the above-described work is extended to cover an area of almost 190,000 km^2 at a snowdrift-permitting scale. A set of observations at multiple scales are used to diagnose the model. These range from point observations, snow surveys, and lidar observations at a catchment scale, and indices derived from remotely sensed imagery at the mountain range scale. The inclusion of wind-induced and gravitational snow redistribution processes for mountain snowpack forecasts are required to accurately predict the snow mass heterogeneity and the evolution of snow mass and persistence with elevation. Without redistribution, snowpacks are overestimated at high elevations and the transport of snow from high elevation to low elevation is missed. These late-lying, low elevation snowpacks are important for estimating spring runoff. Lastly, the technical challenges of running snowdrift-permitting models over large extents are discussed.

[1] Centre for Hydrology, University of Saskatchewan, Saskatoon, SK, Canada
[2] Global Institute for Water Security, University of Saskatchewan, Saskatoon, SK, Canada
[3] Environmental Numerical Research Prediction, Environment and Climate Change Canada, Dorval, QC, Canada
[4] Numerical Simulation Lab, University of Saskatchewan, Saskatoon, SK, Canada

MODIS does not Capture the Spatial Heterogeneity of Snow Cover Induced by Solar Radiation

CHRISTOPHE KINNARD[1], HAFSA BOUAMRI[2], ABDELGHANI BOUDHAR[2,3], SIMON GASCOIN[4], LAHOUCINE HANICH[3,5], AND ABDELGHANI CHEHBOUNI[3,4]

ABSTRACT

Estimating snowmelt in semi-arid mountain ranges is an important but challenging task, due to the large spatial variability of the snow cover and scarcity of field observations. Adding solar radiation as a snowmelt predictor within empirical snow models is often done to account for topographically-induced variations in melt rates. Here we examine the added value of including different treatments of solar radiation within empirical snowmelt models and benchmarks their performance against MODIS snow cover area (SCA) maps over a 14-year (2003-2016) period. Three spatially distributed, enhanced temperature index models, which respectively include the potential clear-sky direct radiation, the incoming solar radiation, and net solar radiation, were compared with a classical temperature-index (TI) model to simulate snowmelt, SWE, and SCA within the Rheraya basin in the Moroccan High Atlas Range. The enhanced models, particularly the one including net solar radiation, were found to better explain the observed SCA variability compared to the TI model. However, differences in model performance in simulating basin wide SWE and SCA were small. This occurs because topographically-induced variations in melt rates simulated by the enhanced models tend to average out, a situation favored by the rather uniform distribution of slope aspects in the basin. While the enhanced models simulated more heterogeneous snow cover conditions, aggregating the simulated SCA from the 100 m model resolution towards the MODIS resolution (500 m) suppresses key spatial variability related to solar radiation, which attenuates the differences between the TI and the radiative models. Our findings call for caution when using MODIS for calibration and validation of spatially distributed snow models.

---

[1] Département des Science de L'environnement, Centre de Recherche sur les Interactions Bassins Versants – Écosystèmes Aquatiques (RIVE), Université du Québec à Trois-Rivières, Trois-Rivières, QC, Canada

[2] Water Resources Management, Valorisation and Remote Sensing Team, Faculty of Sciences and Technology, Sultan Moulay Slimane University, Beni Mellal, Morocco

[3] Center for Remote Sensing Application (CRSA), Mohammed VI Polytechnic University, Ben Guerir, Morocco

[4] CNRS/CNES/IRD/UPS, Centre d'Études Spatiales de la Biosphère (CESBIO), University of Toulouse, Toulouse, France

[5] L3G Laboratory, Earth Science Department, Faculty of Sciences and Techniques, Cadi Ayyad University, Marrakech, Morocco

Towards the Incorporation of Adaptive Viewing in Observing System Simulation Experiments (OSSEs) to Preferentially View Snow-Covered Terrain

COLIN P. MCLAUGHLIN[1], BARTON A. FORMAN[1], AND LIZHAO WANG[1]

ABSTRACT

Observing system simulation experiments (OSSEs) can be used to develop optimal satellite configurations to help improve the characterization of snow mass, and hence, global freshwater. Results from OSSEs are dependent on the quality of the geophysical retrievals obtained, which are a function of the spacecraft platform architecture, the number of spacecraft in a given constellation, and the types of satellite-based sensors used. Typically, retrievals are acquired using a static (fixed) look angle. This project explores how a dynamic look angle through the usage of adaptive viewing can be used to collect more data of specific targets of interest (e.g., snow-covered mountain peak rather than a snow-free valley) in order to improve the end results of an OSSE and thereby enhance our understanding of hydrologic phenomena.

Adaptive viewing is the process of preferentially viewing an area of interest by changing a satellite's field of view as a function of look angle. Adaptive viewing can help increase the snow-covered land that is viewed, thereby improving predictions of snow-water equivalent and associated surface runoff. In this project, a random walk algorithm with re-initialization is used to determine an optimal viewing strategy as a function of user-defined satellite parameters. To quantify the marginal benefit of an adaptive viewing strategy using the random walk approach over fixed viewing, toy experiments are first performed using a synthetically-generated binary surface target to model snow-covered land versus snow-free land. The toy experiments include sensor constraints such as a maximum look angle and maximum rate-of-change in look angle to better diagnose the marginal improvements associated with adaptive viewing relative to static viewing. Lessons learned from the toy problem experiments are then applied across the globe using real-world remote sensing datasets to maximize viewing of snow-covered land, topographical elevation, change in forest density, and irrigated agricultural areas.

[1] University of Maryland, College Park, MD, USA

Improving Microwave Volume Scattering Based SWE Retrieval Performance using SnowEx 2017 SnowSAR Observations

DO-HYUK KANG[1], JIYUE ZHU[2], EDWARD KIM[1,3], AND LEUNG TSANG[2]

ABSTRACT

Numerous efforts have been recently made geared toward a satellite mission to observe SWE from space. This paper evaluates the performance of the previously developed active microwave volume scattering algorithm to retrieve SWE. Validation is explored by comparing measurements of the backscattering coefficient ($\sigma_{o},$) of snow with *in situ* snow observations including snow grain size, temperature, density, and depth which are direct contributors to microwave signatures.

Specifically, we used volume scattering observations from the ESA (European Space Agency) airborne SnowSAR during SnowEx 2017 along with *in situ* snowpit measurements. X- and Ku-band SnowSAR observations are compared with a forward radiative transfer model such as DMRT-Bic. Comparison statistics of retrieved SWE and observed SWE will be reported. In addition, results from sensitivity studies using *in situ* and modeled snow observations are demonstrated to evaluate the influence of snow microstructures and the background contribution of the lower boundary of the snowpack on the microwave volume scattering signals.

[1] ESSIC, University of Maryland, College Park, MD, USA
[2] Radiation Laboratory, Department of Electrical Engineering and Computer Science, The University of Michigan, Ann Arbor, MI, USA
[3] NASA Goddard Space Flight Center, Greenbelt, MD, USA

Comparison of NASA MODIS / VIIRS Cloud-Gap-Filled with other Satellite-Derived Snow-Cover Maps

DOROTHY K. HALL[1], GEORGE A. RIGGS[2], NICOLO E. DIGIROLAMO[2], ANGELA M. ERB[3], AND CRYSTAL B. SCHAAF[3]

INTRODUCTION

Seasonal snow cover regulates the surface energy balance, in large part due to its high albedo and low thermal conductivity, influencing earth-surface processes and availability of water resources. Since 2000 NASA standard snow-cover maps have been available from the MODerate-resolution Imaging Spectroradiometer (MODIS) on the Terra satellite and, since 2002, on the Aqua satellite at a spatial resolution up to 500 m, and since 2012 from the Suomi-National Polar Program (S-NPP) Visible Infrared Imaging Radiometer Suite (VIIRS) at 375-m resolution. These daily, global snow-cover products are used extensively in studies of seasonal snow-cover dynamics at regional and hemispheric scales. A limitation of these products is that they cannot map snow through cloud cover. To mitigate this limitation, NASA cloud-gap-filled (CGF) standard snow-cover map products were developed from the Terra/Aqua MODIS and the S-NPP VIIRS sensors and are named MOD/MYD10A1F and VNP10A1F, respectively.

The purpose of this work is to evaluate the accuracy of the NASA CGF snow-cover maps through comparisons with existing satellite-derived snow-cover maps. *First*, we look at time series data to assess continuity between the NASA CGF snow products. We then compare VNP10A1F with NOAA's Interactive Multisensor Snow and Ice Mapping System (IMS) 1- and 4-km snow maps (Helfrich *et al.*, 2007; Woods and Helfrich, 2019) for a 2,487,610 km^2 study area in the western U.S. and southern Canada (Figure 1). *Next,* we compare VNP10A1F snow maps with snow maps derived from 30-m resolution USGS Landsat-8 Operational Land Imager (OLI) data using the USGS fSCA (Selkowitz and Forster, 2016) and the Landsat CFMask-derived snow map. The CFMask is the USGS operational cloud and cloud shadow detection algorithm that is applicable with data from Landsat and Sentinel-2 (Zhu and Woodcock, 2012; Foga *et al.*, 2017). For the comparisons using fSCA, which provides fractional snow cover, we used a cutoff to create binary snow maps where snow-cover fraction ≥15% in a pixel is mapped as "snow," and everything else is mapped as "other."

RESULTS

Comparison of NASA CGF snow products

Previous results show excellent correspondence between the Terra MODIS CGF snow maps (MOD10A1F) and the S-NPP CGF snow maps (VNP10A1F), with small differences in snow-cover extent (SCE) attributed to differences in spectral bands, native resolution, sun angle and time of day of image acquisition (Riggs and Hall, 2020). A high degree of continuity is expected because the VNP10A1F algorithm was developed specifically to match the MOD/MYD10A1F algorithm to

[1] Earth System Science Interdisciplinary Center / University of Maryland, College Park, MD, USA
[2] Science Systems and Applications Inc., Lanham, MD, USA
[3] School for the Environment, University of Massachusetts Boston, Boston, MA, USA

enable the creation of a climate data record (Hall *et al.*, 2019; Riggs *et al.*, 2019). Figure 2 shows an example of the excellent agreement (R = 0.099, mean bias -0.0001 million km^2) between SCE measured using the MOD10A1F and VNP10A1F for the Upper Klamath Lake basin which is within the study area shown in Figure 1.

Figure 1. Study area (2,487,610 km^2) covering all or parts of 11 states in the western U.S. and part of southern Canada.

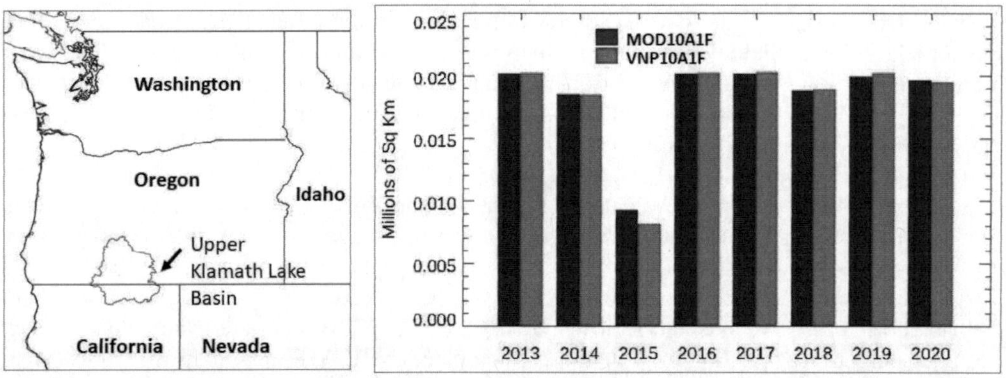

Figure 2. Location map is on the left. Graph shows winter (DJF) snow extent in millions on km^2 in the UKL Basin as measured using the MOD10A1F and VNP10A1F NASA cloud-gap-filled snow-cover products.

Comparison of NASA CGF snow products with other satellite-derived snow products

To assess the agreement of the NASA CGF daily snow maps with other established SCE maps, we plotted the SCE in km^2 from the VNP10A1F and IMS 1-km and 4-km resolution SCE maps for water years (WY) 2016, 2017 and 2018. Results from WY 2018 are shown in Figure 3. In all three years, the IMS 1- and 4-km results are virtually identical (R^2 = 99), with a mean bias of -0.002 km^2 (for 2016) and -0.004 (for 2017 and 2018) (Table 1).

The coefficient of determination, R^2, between the VIIRS CGF and IMS snow products is high in all three years and, for example, in the 2018 WY, only 4 percent of the days on which data were available between the first of October and the end of May show large (≥15 percent) differences between the VNP10A1F and IMS 1-km SCE (Table 1). However, there are many days in different areas within the study area (see Figure 1) where there are large differences in the SCE mapped using VIIRS and IMS, for example in early November of 2017 and at the end of January/early February of 2018 (Figure 3). There are two main causes for these differences. 1) The IMS often tends to "fill in" and thus map too much snow, probably due to analysts' decisions to fill in partially snow-covered areas, and because of the IMS coarser resolution compared to the NASA CGF snow maps.

2) The VIIRS CGF maps cannot "see" new snow if it falls while the area is cloudy, or "see" existing snow that has melted while clouds are present.

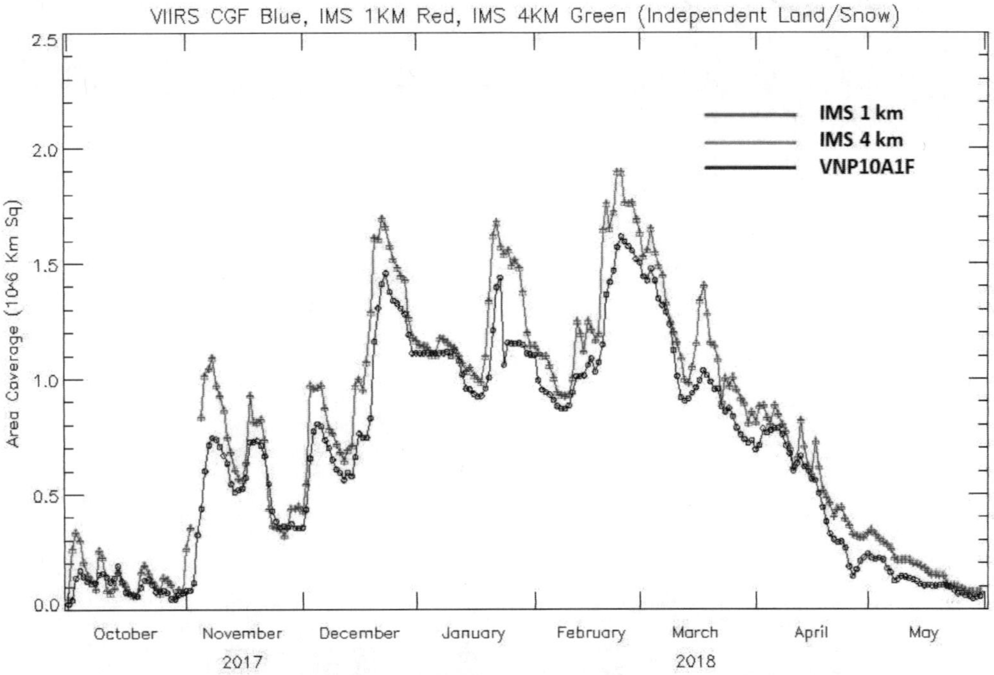

Figure 3. IMS 1- and 4-km and VNP10A1F SCE in km2 in the western U.S./southern Canada study area for WY 2018.

Table 1. Comparison of time series snow maps (VNP10A1F minus IMS 1 km) for three partial water years in the northwestern U.S./southern Canada study area shown in Figure 1.

Water Year	RMSE*	R^2	Mean Bias*	Percent of days for which the difference exceeds 15%
2016	0.166	0.962	0.102	5
2017	0.143	0.970	0.080	3
2018	0.160	0.963	0.119	4

* unit is = 1,000,000 km^2

Comparison of NASA CGF snow products with higher-resolution Landsat-derived snow products

To illustrate the first point indicating that the IMS maps tend to "fill in" and thus sometimes map too much snow, we compare VIIRS and IMS snow maps with higher-resolution (30 m) Landsat OLI data for a 29 November 2015 Landsat scene (path/row 039/026) located at the U.S./Canada border in northcentral Montana and southeastern Alberta/southwestern Saskatchewan (Figure 4). The VIIRS map provides greater detail than does IMS. In the VIIRS map, Figure 4c, the various shades of purple and white represent different values of the normalized-difference snow index (NDSI) that can be used to estimate fractional snow cover within a pixel, with the white pixels indicating nearly 100 percent snow coverage and the darkest purple pixels indicating ~10 percent.

In Figure 5 we show difference maps using VNP10A1F vs the USGS fSCA snow map, and IMS 1-km *vs* fSCA for the same general area in northern Montana / southern Canada on 29 November 2015 as shown in Figure 4. To develop these snow maps, two Landsat-8 OLI scenes were stitched together. Both the VNP10A1F VIIRS and IMS maps agree well with the Landsat fSCA map, with the VNP10A1F *vs* fSCA agreement being somewhat higher, at 90 percent, than the IMS 1-km *vs* fSCA agreement at 88 percent.

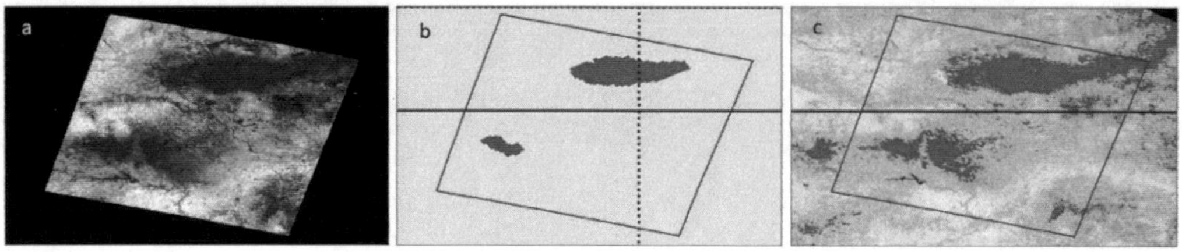

Figure 4. The study area spans the U.S./Canada border in northcentral Montana and southern Canada, 29 November 2015; all are mapped to a common lat/long grid at 500-m resolution. a) Landsat-8 OLI bands 4, 3, 2 in which snow is white and non-snow features are black (i.d. #LC08_L2SP_039026_20151129_20200908_02_T1); b) IMS snow map in which snow is light purple and non-snow features are green; c) VIIRS VNP10A1F snow map in which the various shades of purple and white represent different NDSI values; white is nearly 100 percent snow cover and dark purple is ~10 percent snow cover.

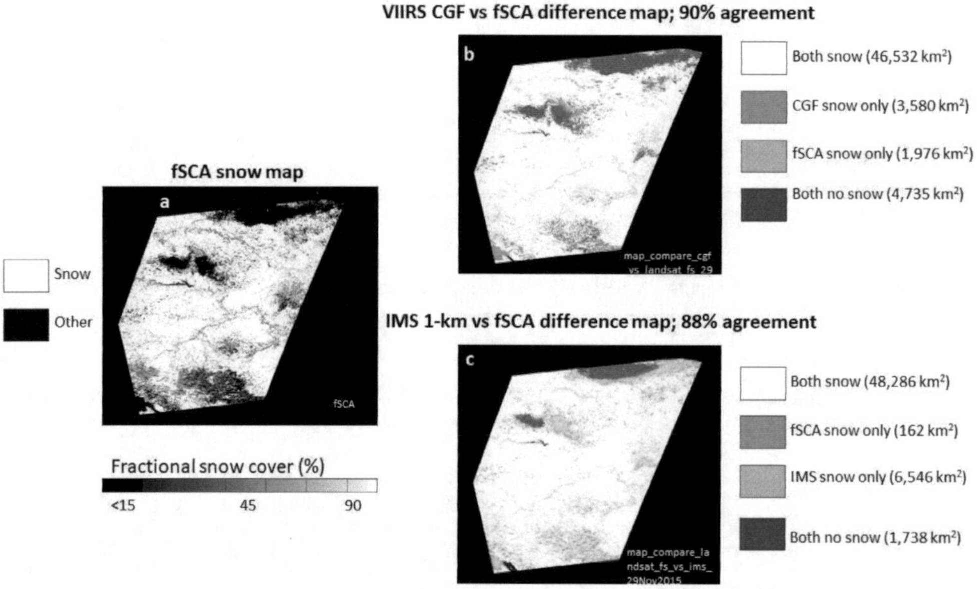

Figure 5. 29 November 2015 maps of an area in northern Montana/southern Canada mapped to a common lat/long grid at 500-m resolution. a) Landsat-8 OLI USGS fSCA snow map; b) VNP10A1F *vs* fSCA snow map shows 90 percent agreement; c) IMS 1-km snow map *vs* fSCA snow map shows 88 percent agreement. The following Landsat-8 OLI scenes were used to develop the snow maps in these figures: LC08_L2SP_039026_20151129_20200908_02_T1 and LC08_L2SP_039027_20151129_20200908_02_T1.

We also looked at 12 March 2021 Landsat-8 OLI scenes from the study area in southern Oregon/northern California that includes the UKL basin. We compared the Landsat fSCA snow map with the CFMask snow map, finding 90 percent agreement; we also compared VNP10A1F *vs* CFMask and found 87 percent agreement. Then we compared VNP10A1F *vs* fSCA for the same area, finding 91 percent agreement (Figure 6).

Uncertainties in cloud-gap-filling

The NASA fully-automated standard MODIS/VIIRS CGF maps can miss snow during cloudy conditions because a snow-cover decision cannot be updated until the clouds clear, thus the snow-map result from MOD10A1 from the most recent previous clear day, is retained in the MOD10A1F (CGF) algorithm. (MOD10A1 is the NASA standard daily snow map that includes clouds and is input to the MOD10A1F algorithm.) However, the IMS map production is not fully automated and thus key information such as from meteorological station data and local observations is used in the production of their daily maps, so the IMS products can be advantageous during extended periods of cloud cover. In Figure 7 the IMS map (lower panel) correctly captures some snow cover beneath clouds that is missed by the MOD10A1 algorithm (upper panel). The reason was investigated, and it was determined that there was an error in the MOD35 cloud mask used as input to the MOD10A1 snow-mapping algorithm. The shape of the 'cloud' feature on this 25 March 2012 MOD10A1F snow map shown in the red rectangular area is similar to a snow feature in the NOAA IMS 4-km snow map from the same day (lower panel). When the cloud mask was removed from the MOD10A1 daily snow map, a feature that appears to be snow is revealed (not shown). The determination that this really is a snow feature was confirmed by visual interpretation of other satellite data and NOAA station data.

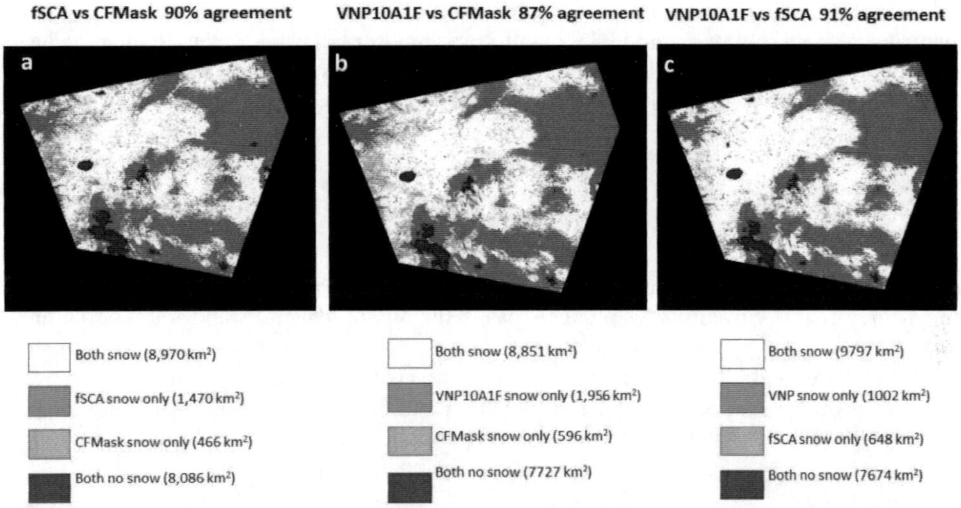

Figure 6. Comparison of Landsat-derived and VNP10A1F snow maps for an area in southern Oregon and northern California that includes the UKL basin, 12 Marsh 2021, mapped to a common lat/long grid at 500-m resolution. a) Landsat-8 OLI USGS fSCA snow map *vs* CFMask snow map; b) VNP10A1F snow map vs Landsat CFMask snow map; c) VNP10A1F snow map *vs* Landsat OLI fSCA snow map.

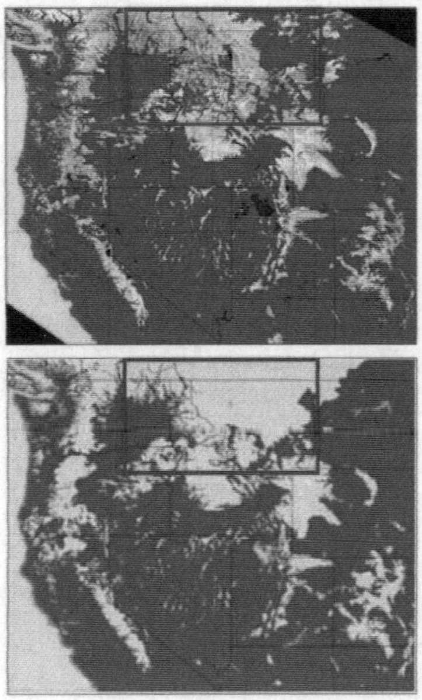

Figure 7. Due to cloud obscuration in the MOD10A1F product which is input to MOD10A1F, the MODIS CGF map (upper panel) does not show as much snow in Montana (within the red rectangle) as was mapped by IMS (lower panel) and was visible on other satellite imagery and meteorological station data on these 25 March 2012 snow maps.

DISCUSSION AND CONCLUSION

Under clear-sky conditions, differences in native resolution, sun angle and time of day account for most of the differences between the NASA and NOAA snow maps (Riggs and Hall, 2020). However, time series show that the IMS snow maps often tend to map more snow compared to the MODIS and VIIRS. When there are several days of cloud cover in a row, the NASA CGF maps may underperform the IMS maps on a given day in an area where a snowstorm produces a large snowfall while the area is cloudy or snow may melt (and thus cannot be updated by the CGF algorithms). Within the MOD10A1F algorithm there is a cloud persistence data layer that informs the user of the number of days since the last clear day when an observation was made.

The NASA CGF (in this case VNP10A1F) and IMS snow maps show 90 and 88 percent agreement, respectively, with the fSCA map for an area in northern Montana/southern Canada on 29 November 2015 (Figure 5). For an area in southern Oregon/northern California, the agreement is similar on 12 March 2021 (Figure 6). Many additional comparisons with Landsat-derived maps under both clear and partly-cloudy conditions are needed to gain quantitative insights into this analysis of NASA CGF uncertainties. This preliminary study provides examples of some ways that the NASA CGF uncertainties can be evaluated.

The NASA standard MODIS (MOD/MYD1A1) and VIIRS (VNP10A1) snow-cover maps have uncertainties that are consistent and have been well described in the peer-reviewed literature. Though there is no current snow product that is the 'gold standard,' and thus the community-accepted 'ground truth' for snow mapping, when the NASA CGF snow-map products (that use MOD/MYD10A1 and VNP10A1 as input) were compared with IMS 1- and 4-km snow maps and 30-m resolution Landsat-derived snow maps (fSCA and CFMask) for this small study, the NASA CGF standard SCE maps, MOD/MYD10A1F and VNP10A1F, consistently show good results.

Data availability

The NASA standard Terra and Aqua MODIS cloud-gap-filled snow-cover products may be found at: *https://nsidc.org/data/MOD10A1F/versions/61/print* and *https://nsidc.org/data/MYD10A1F/versions/61*, respectively. The NOAA IMS 1- and 4-km resolution snow maps are available at: *https://nsidc.org/data/g02156*. Landsat Operational Land Scanner (OLI) data are available from the EarthExplorer site at *https://earthexplorer.usgs.gov/*.

Acknowledgement. This research was supported by the NASA Terrestrial Hydrology Program, grant number 80NSSC18K1690.

REFERENCES

Foga S, Scaramuzza PL, Guo S, Zhu Z, Dilley Jr RD, Beckmann T, Schmidt GL, Swyer JL, Hughes MJ, Laue B. 2017. Cloud detection algorithm comparison and validation for operational Landsat data products. *Remote Sensing of Environment*, **194**: 379–390.

Hall DK, Riggs GA, DiGirolamo NE, Román MO. 2019. Evaluation of MODIS and VIIRS cloud-gap-filled snow-cover products for production of an Earth science data record. *Hydrology and Earth System Science*, **23**: 5227–5241. *https://doi:10.5194/hess-23-5227-2019.*

Helfrich SR, McNamara D, Ramsey BH, Baldwin T, Kasheta T. 2007. Enhancements to, and forthcoming developments in the Interactive Multisensor Snow and Ice Mapping System (IMS). *Hydrological Processes,* **21**(12): 1576-1586.

Riggs GA, Hall DK, Román MO. 2017. Overview of NASA's MODIS and Visible Infrared Imaging Radiometer Suite (VIIRS) snow-cover Earth System Data Records. *Earth System Science Data*, **9**: 765–777. https://doi:10.5194/essd-9-765-2017.

Riggs GA, Hall DK. 2020. Continuity of MODIS and VIIRS Snow Cover Extent Data Products for Development of an Earth Science Data Record. *Remote Sensing*, **12:** 3781. https://doi:10.3390/rs12223781.

Riggs GA, Hall DK, Román MO. 2019. MODIS snow products user guide for Collection 6.1 (C6.1), available at: https://modis-snow-ice.gsfc.nasa.gov/?c=userguides.
Selkowitz DJ, Forster RR. 2016. Automated mapping of persistent ice and snow cover across the western U.S. with Landsat. *Journal of Photogrammetry and Remote Sensing*, **117**: 126–140. https://dx.doi.org/10.1016/j.isprsjprs.2016.04.001.
Woods J, Helfrich S. 2019. Interactive Multisensor snow and ice Mapping System (IMS) Upgrades and Improvements. *Proceedings of the 76th Eastern Snow Conference*, Fairlee, Vermont.
Zhu Z, Woodcock CE. 2012. Object-based cloud and cloud shadow detection in Landsat imagery. *Remote Sensing of Environment,* **118**: 83-94. https://doi.10.1016/j.rse.2011.10.028.

Snow Satellite Mission Concept Considerations, Key Questions, and Needed Tools

EDWARD KIM[1], PAUL HOUSER[2], AND ANA BARROS[3]

ABSTRACT

In the recent Earth Science Decadal Survey, the Water Panel expressed, as an overarching goal, an integrated system of global water cycle-related observations and models. Such a system would provide a better picture of the global state of the water cycle (storage terms as well as fluxes), enabling advancements in science and applications. Seasonal snow (specifically snow depth and/or water equivalent) is listed in the Explorer category of 7 important future measurements (not all of which are water-related) because it was seen as a critical quantity where improvement is needed as well as quantity for which observing options were rated as being "mature enough".

"Global" products of seasonal snow, however, already exist, and certain satellite observations already exist. So, any viable snow mission concept must improve upon the existing products. And, while many observing techniques are sensitive to snow depth and/or to SWE, all have limitations. There are a variety of science targets for a future snow satellite mission, and perhaps a wider range of target applications. How to assess the gap between what current snow products provide and the needs to address desired science or applications, and to compare that with what observation and model capabilities might be able to provide is the topic of this paper. We will present recent thoughts on how to make the above gap assessments, how to compare the gaps to observation and model capabilities, and we will describe some of the tools that will be needed along this road to help define and guide a snow mission concept.

[1] NASA Goddard Space Flight Center, Greenbelt, MD, USA
[2] George Mason University, Fairfax, VA, USA
[3] University of Illinois at Urban-Champaign, Urbana, IL, USA

Scattering Mechanics of Freshwater Ice Derived Through Polarimetric Decomposition from Sledborne Scatterometers

G.E. GUNN[1], A. THOMPSON[1], AND J. FERGUSON[1]

ABSTRACT

From the outset of radar observation of geophysical variables, studies investigating time series of radar backscatter from freshwater ice have hypothesized that the timing and regional distribution of high backscatter was caused by vertical columnar bubbles encapsulated within the ice. Bubbles would cause coherent scatter in the forward direction, followed by specular reflection off the high dielectric-contrast ice-water interface and back to the sensor (dubbed "double-bounce"). Technological improvements have allowed for the acquisition of the full scattering matrix, allowing for the calculation of co-polarized phase differences, and the application of polarimetric decomposition algorithms that separate the signal into surface, double bounce and volume scatter interactions. Spaceborne acquisitions using a range of frequencies from L- to X-band (1.2 to 9.6 GHz) have revealed the dominant scattering mechanism to be surface bounce regardless of the timing of observation (early *vs.* late winter), however the understanding of the physical properties of the ice is limited by the spatial resolution of synthetic aperture radar (SAR) acquisitions.

A time-series of co-pol phase difference and scattering mechanisms were derived from fully polarimetric backscatter collected at X- and Ku-band (9.6, 17.2 GHz, respectively) using the University of Waterloo Scatterometer (UWScat) on Malcolm Ramsey Lake near Churchill, MB. Interactions near the snow-ice and ice-water interfaces are able to be separated during processing and can be combined for comparison with spaceborne observations. Single bounce was the overall dominant signal interaction for both interfaces at incidence angles typically observed by SAR (e.g. 27 - 54°). UWScat backscatter was positively correlated to increases in bubble density, however the polarimetric parameters consistently indicate the source of backscatter to be roughness at the ice-water interface. This study contributes to the emerging body of evidence that supports the hypothesis that bubble density serves to increase height deviations at the ice-water interface sufficient to appear rough at wavelengths commonly used on spaceborne SAR.

[1] Department of Geography and Environmental Management, University of Waterloo, Ontario, Canada

Measuring Changes in Snowpack SWE Continuously on a Landscape Scale using Lake Water Pressure

HAMISH D. PRITCHARD[1], DANIEL FARINOTTI[2], AND STEVEN COLWELL[1]

ABSTRACT

The seasonal snowpack is a globally important water resource that is notoriously difficult to measure. Existing instruments make measurements of falling or accumulating snow water equivalent (SWE) that are susceptible to bias, and most can represent only a point in the landscape. Furthermore, the global array of SWE sensors is too sparse and too poorly distributed to be an adequate constraint on snow in weather and climate models. We present a new approach to monitoring snowpack SWE from time series of lake water pressure. We tested our method in the lowland Finnish Arctic and in an alpine valley and high-mountain cirque in Switzerland, and found that we could measure changes in SWE and their uncertainty through snowfalls with little bias and with an uncertainty comparable to or better than that achievable by other instruments. More importantly, our method inherently senses change over the whole lake surface which can be several square kilometers, or hundreds of millions of times larger than the aperture of a pluviometer. This large scale makes our measurements directly comparable to the grid cells of weather and climate models. We find, for example, snowfall biases of up to 100% in operational forecast models AROME-Arctic and COSMO-1. Seasonally-frozen lakes are widely distributed at high latitudes and are particularly common in mountain ranges, hence our new method is particularly well suited to the widespread, autonomous monitoring of snow-water resources in remote areas that are largely unmonitored today. This is potentially transformative in reducing uncertainty in regional precipitation and runoff in seasonally-cold climates.

[1] British Antarctic Survey, Cambridge, UK
[2] Swiss Federal Institute for Forest, Snow and Landscape Research WSL, Birmensdorf, Switzerland

History of Winter Carnival Events in College Archives and Snowfall Observations in Williamstown, MA, 1913-2010

HAYDEN GILLOOLY[1] AND ALICE BRADLEY[1]

ABSTRACT

In areas where winter temperatures often hover around freezing, annual snow accumulation totals can be highly variable, with slightly colder winters having high total accumulation and slightly warmer winters bringing more rain. A long-term record of snow accumulation in the area around Williamstown, Massachusetts (USA) dating back to 1913 documents that variability, with observed monthly accumulation totals for most years and annual accumulation totals during the monthly data gaps. Decadal mean snowfall totals dropped around the 1970s, with mean accumulation of 89.69 cm annually before the 1970s, compared to 70.56 cm after 1970. The bulk of this difference comes early in the snow season, with dramatically lower snow accumulation recorded in October through December during the last 50 years. Records from the Williams College archives, including issues of the student newspaper dating back to the late 1800s, provide context for these observations. Of particular interest are the activities around Winter Carnival, a mid-February event that has historically included building snow sculptures, skiing, and other snow-based celebrations but that in recent years has often been snow-free. For a region that depends in part on winter tourism, the increasing unreliability of winter snow cover has economic impacts, in addition to hydrological and ecological impacts.

[1] Geosciences Department, Williams College, Williamstown, MA, USA

Comparison of *in situ* Snow Depth Measurements and Impacts on Validation of Unpiloted Aerial System Lidar over a Mixed-Use Temperate Forest Landscape: A Case Study in Durham, New Hampshire, United States

HOLLY PROULX[1], ELIZABETH A. BURAKOWSKI[2], ADAM G. HUNSAKER[1,2], JENNIFER M. JACOBS[1,2], FRANKLIN B. SULLIVAN[2], MICHAEL PALACE[2,3], EUNSANG CHO[1,2], AND CAMERON WAGNER[1]

ABSTRACT

The accuracy of snow depth measurements depends on the measuring device and the conditions of the site and snowpack in which it is being used. This research informs practices to consider when validating unpiloted aerial system (UAS) lidar-derived snow depth measurements in shallow snowpacks. We conducted three snow depth sampling campaigns from December 2020 to March 2021 that included three transects from an open field to coniferous, mixed, and deciduous forest plots at Thompson Farm in Durham, NH. Snow depth was measured using a magnaprobe, a snow tube, field cameras, and UAS-lidar. Depth measurements were compared by measurement technique and land type. Impacts of the leaf litter on snow depth measurements were also quantified.

The open field snow depths (16.6 cm ± 0.5) were consistently deeper than the forested area depths (13 cm ± 0.5). Over the three sampling campaigns, there was nearly a 20% difference between the magnaprobe (15 cm) and snow tube (12.7 cm) average snow depths, with a greater difference in the forest than the field. This was likely due to the tube encountering woody debris and the magnaprobe over-probing through the leaf litter and possibly into the soil; over-probing was mitigated by creating a new cross-like shaped tip to limit probing into the litter layer, which yielded snow depth more consistent with snow tube measurements. The snow depth measurements estimated from visual inspection of snow stake photos routinely taken by the field cameras had approximately 1 cm precision. The results indicate that in shallow snowpacks, snow depth estimates are significantly different among instruments and those difference vary by land use and local surface conditions (i.e. vegetation height, leaf litter, and woody debris).

[1] Department of Civil and Environmental Engineering, University of New Hampshire, Durham, NH USA

[2] Earth Systems Research Center, Institute for the Study of Earth, Oceans, and Space, University of New Hampshire, Durham, NH, USA

[3] Department of Earth Sciences, University of New Hampshire, Durham, NH, USA

NASA SnowEx 2020 and 2021 Campaigns in the Western U.S.

HP MARSHALL[1], CARRIE VUYOVICH[2], CHRIS HIEMSTRA[3], KELLY ELDER[4], MICHAEL DURAND[5], AND ELIAS J. DEEB[6]

ABSTRACT

NASA SnowEx's overarching objective is to measure and monitor snowpack properties from space, in particular snow water equivalent and albedo. To achieve this, a series of coincident field, airborne, and modeling campaigns have been initiated. These campaigns are testing a wide range of snow remote sensing technologies, with a focus on a pathway to eventual spaceborne deployment. The NASA snow remote-sensing community has agreed that no single sensor or approach solves the spaceborne snow monitoring problem; therefore, we are testing approaches that combine multiple remote sensing products with field observations within the framework of land surface models. SnowEx 2017 focused efforts on two western Colorado sites: Grand Mesa and nearby Senator Beck Basin, targeting impacts of forest cover on remote sensing approaches. SnowEx 2020 was spatially and temporally broader. It linked a second Intensive Observation Period (IOP) at Grand Mesa with an L-band InSAR time-series experiment for thirteen locations scattered across the Contiguous United States' western half. SnowEx 2021 continued the L-band time series, expanded efforts into northern prairie/agricultural environments, and collected novel albedo measurements. SnowEx 2022 planning is underway, with the campaign moving to Alaska to study various remote sensing technologies in tundra and taiga environments, and the impact of permafrost on radar retrieval.

[1] Department of Geosciences, Boise State University, Boise, ID, USA
[2] Hydrological Sciences Laboratory, NASA Goddard Space Flight Center, Greenbelt, MD, USA
[3] US Army Corps of Engineers Cold Regions Research and Engineering Laboratory, Ft. Wainwright, AK, USA
[4] US Forest Service, Fort Collins, CO, USA
[5] School of Earth Science, The Ohio State University, Columbus, OH, USA
[6] US Army Corps of Engineers Cold Regions Research and Engineering Laboratory, Hanover, NH, USA

Toward Constraining Mountain Stream Flow Constituents by Combining Citizen Scientist Acquired Geochemical Tracers with Sentinel-1 SAR Time Series in Pakistan

JEWELL LUND[1,2,3], RICHARD R. FORSTER[1], ELIAS J. DEEB[2], SUMMER B. RUPPER[1], STEVEN J. BURIAN[3,4], YUSUF JAMEEL[3], HP MARSHALL[5], GHULAM HUSSAIN DARS[3,6], MASOOD ALI[7], ABDUL GHAFOOR[8], AND AZHAR ZAHEER[3]

ABSTRACT

Snow and ice melt are both critical components of streamflow in the transboundary Indus River Basin, which flows through the Hindu Kush, Western Himalaya, and Karakoram mountain ranges and serves as a primary water source for nearly 300 million people in Tibet, India, Afghanistan, and Pakistan. A lack of field measurement and data sharing limits specific understanding of the role of snow and glacier melt in Indus River flows, proliferating uncertainties for anticipating future water availability in a region vulnerable to climate change. Satellite-based synthetic aperture radar (SAR) instruments image day or night and also penetrate cloud cover. Due to the high permittivity of liquid water, C-band (5.5 cm wavelength) SAR imagery can be used to reliably detect meltwater present in the snowpack. This offers spatially explicit information about which areas of the snowpack can potentially contribute to streamflow, and also provides an alternative method for calculating snow cover area that is not subject to the same constraints prevalent in optical satellite imagery. Geochemical tracers offer a means to constrain streamflow constituent estimates (e.g. snowmelt, glacier melt, groundwater, precipitation), and have been used successfully in many temperate watersheds. Their application to glaciated alpine watersheds has been explored, though spatial and temporal isotopic variability in endmembers yields many open questions.

In this study, we analyze geochemical tracers from samples collected by local citizen scientists from 2018-2019 in the heavily glaciated Shimshal watershed in the Karakoram Mountains of Pakistan. We integrate geochemical data with European Space Agency Sentinel-1 SAR imagery in an effort to better constrain streamflow constituents in this remote basin that serves as a headwater to the Indus River in Pakistan. We compare our findings with modeled results from the region, explore the role and potential for citizen science to support climate-related studies, and identify important factors for future work.

---

[1] Department of Geography, University of Utah, Salt Lake City, UT, USA
[2] US Army Corps of Engineers Cold Regions Research and Engineering Laboratory, Hanover, NH, USA
[3] US-Pakistan Center for Advanced Studies in Water
[4] Department of Civil and Environmental Engineering, University of Utah, Salt Lake City, UT, USA
[5] Department of Geoscience, Boise State University, Boise, ID, USA
[6] Mehran University of Engineering and Technology, Jamshoro, Pakistan
[7] Shimshal Nature Trust, Shimshal, Pakistan
[8] Higher Ground Expeditions, Islamabad, Pakistan

Volume Determination and Area-Volume Scaling on a Small Bolivian Cirque Glacier, Charquini SE

J.L. KINCAID[1], I.D. DOBREVA[1], AND A.G. KLEIN[1]

ABSTRACT

Tropical glaciers serve as important climate indicators, as well as crucial water resources for local rural populations and nearby urban population centers. There is concern that this water resource is disappearing at an accelerating rate due to climate change. Policy makers require a better understanding of the amount of water currently contained in these glaciers, as well as of future projections for continued runoff to maximize their ability to store meltwater from these glaciers and to enact effective public policy measures. This study focuses primarily on the small Charquini SE cirque glacier located in the southern Cordillera Real in Bolivia. During an August 2012 field campaign, differential Global Positioning System (dGPS) and ground penetrating radar (GPR) measurements were collected along transects totaling 3189 meters. At this time, the glacier had a surface area of 2.42 square kilometers which has subsequently been much reduced. Initial analysis of the GPR measurements indicates success in capturing the ice rock interface. Ordinary kriging was used to generate both glacier surface and subsurface topographies. In 2012, based on calculations between these two surfaces, this glacier contained 3,900,000 cubic meters of ice. Historical aerial photos of this glacier, acquired in 1963 and 1983, were also obtained and are being processed to generate surface topographies for previous periods. This will enable volume loss for different periods to be calculated. In addition to calculating volumes directly, area volume scaling relationships will be determined based on our measurements on Charquini SE and other modern and paleoglacier measurements in the Bolivian Andes.

[1] Department of Geography, Texas A&M University, College Station, TX, USA

Evaluating and Improving Northeastern US Snow in the National Water Model by Leveraging Advanced Mesonet Observations: Point Simulations and Sensitivity Experiments

JUSTIN R. MINDER[1], PAT NAPLE[1], AND THEODORE W. LETCHER[2]

ABSTRACT

The NOAA National Water Model (NWM) is used for operational forecast of hydrology, including surface snowpack, across the continental United States (US). It initializes surface snowpack conditions using a simple deterministic analysis cycling every hour and represents snowpack physics using the Noah-MP land surface model. However, performance of the NWM at simulating snow over the northeastern United States is poorly characterized, in part due to limitations in the observational network. Our group is working to improve snow state initialization and prediction within the NWM by leveraging observations from the advanced New York State Mesonet (NYSM; *http://nysmesonet.org/*), which includes 126 stations with detailed meteorology measurements and sub-networks with additional snowpack water equivalent (SWE) and surface energy flux measurements.

As part of this larger effort, we are identifying specific sources of bias in model forcing and physical parameterizations to target for improvement. This presentation will focus on the results of single-column simulations of snowpack properties at NYSM stations from the 2019-2020 and 2020-2021 snow seasons. These simulations are forced by meteorological measurements (temperature, humidity, winds, solar radiation, precipitation) from the NYSM in order to minimize forcing uncertainty and focus on model structural biases. Simulated snow depth and SWE are evaluated against automated NYSM and manual snow course measurements taken around select NYSM stations.

When using the baseline configuration of the NWM, most stations show substantial positive biases in snow depth, SWE, and snow cover duration. To isolate and reduce the sources of these biases, we conduct controlled sensitivity experiments wherein we re-run the simulations after changing a selected model component. Results from these experiments show that large positive biases in simulated snow accumulation are likely caused by a poor representation of precipitation phase partitioning and can be substantially reduced by changing the temperature threshold used or by diagnosing precipitation phase from an operational high-resolution atmospheric model. We also use these results to examine biases in snow albedo that may contribute to biases in springtime snow ablation rates. Preliminary results suggest that reductions in snow albedo following rain-on-snow events are not adequately captured by the model.

[1] University at Albany, State University of New York, Albany, NY, USA
[2] Cold Regions Research and Engineering Laboratory, U.S. Army Engineer Research and Development Center, Hanover, NH, USA

Exploring the History of Snow Research Through the Presentations of the Eastern Snow Conference

KRYSTOPHER J. CHUTKO[1]

INTRODUCTION

In 2021, the Eastern Snow Conference (ESC) enters its eighth decade as a forum for presenting and discussing the wide variety of topics related to snow and ice research while also covering broader topics of winter conditions and seasonally/perennially frozen environments. Over this period, the technology, methodology, and language used to present this research has changed considerably, reflecting the large-scale changes observed in society in general – for example, rapid growth in transport infrastructure, near-Earth space flight, and massive changes in global telecommunications.

Since 1952, the research presented at the Eastern Snow Conference has been documented in the annual Proceedings of the Eastern Snow Conference. The papers, abstracts, and extended abstracts contained within these documents represent a history of the development of the snow and ice sciences over this period of time. In most cases, paper/presentation titles contain keywords necessary to convey the primary type, method, and/or location of research contained within the text of the paper. Therefore, they can be used as a brief summary of research and as a mean to evaluate trends in research via text mining techniques. This study evaluates the changes in words used in the ESC Proceedings to capture a glimpse of the changing approaches to research in snow and ice.

METHODOLOGY

Titles from papers and abstracts presented in the Proceedings of the Annual Eastern Snow Conference were organized by decade, from 1952-60 through 2011-2019 (there was no publication of papers in 1951 or 2020). Hyphenated words (e.g., snow-fall, sub-Arctic) and words commonly spelled in multiple ways (e.g., snow cover and snowcover) were simplified to a common form (e.g., snowfall, subarctic, snowcover). Words identifying specific locations (eg, New Hampshire = newhampshire, Saint John River = saintjohnriver) were similarly simplified. The decadal lists of titles were analyzed in *R* using the "tm" package (Feinerer *et al.*, 2008). At this stage, capitalization, numbers, punctuation, and several words were removed from the lists to simplify the analysis, including common text mining stopwords lists (in both French and English: eg, la, et, par, in, as, and, ...) and words common to the literature used that overwhelmed the analysis (e.g., snow, ice). In total, 9697 words were identified from 1356 paper titles. The average title length increased from 4.9 substantive words in the 1950s to 8.8 substantive words in the 2010s.

Decadal word lists were arranged as word clouds using the "wordcloud" package in *R* (Fellows, 2018). A maximum of 200 distinct words were used in each cloud to minimize cluttering with infrequently used words. Words in the cloud were distinguished by font size and colour based on frequency of use. 35% of words were rotated 90° to facilitate packing of the word cloud.

[1] Department of Geography and Planning, University of Saskatchewan, Saskatoon, SK, Canada

RESULTS AND DISCUSSION

Decadal word clouds are shown in Figure 1. As should be expected across the seven decades analyzed here, words used in paper titles have changed according to the development of the science, techniques and methods used, and the general compartmentalization of research themes. There are several words common throughout, such as "snowmelt", "water", and "forecast", all of which are dominant themes in snow and ice research. However, the frequency of use of some have changed markedly through time. Snow water equivalent, or SWE, is not used in a title until 1970 but becomes dominant in the 21st century. On the other hand, "snowmelt" shows a decreasing trend in use (Fig. 2).

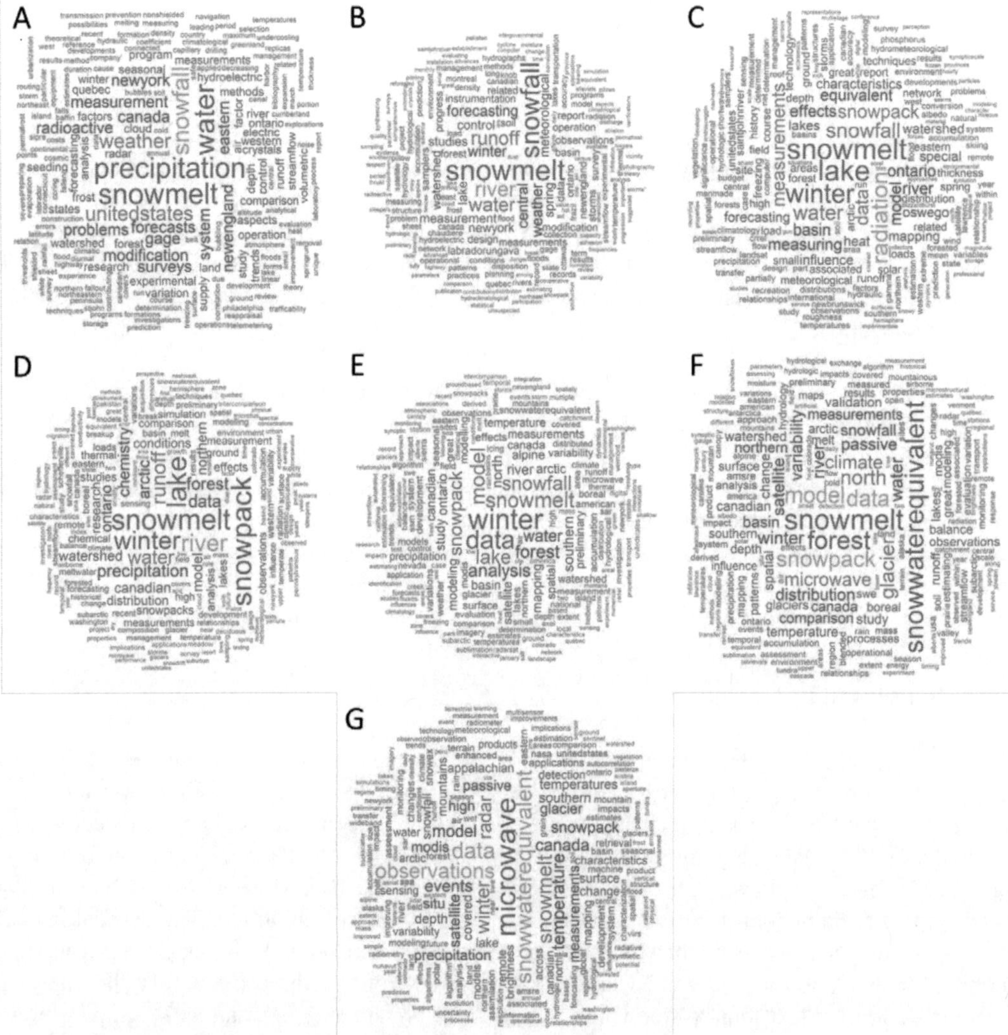

Figure 1. The frequency of words used in Proceedings of the ESC titles for A: 1952-60, B: 1961-1970, C: 1971-1980, D: 1981-1990, E: 1991-2000, F: 2001-2010, and G: 2011-2019.

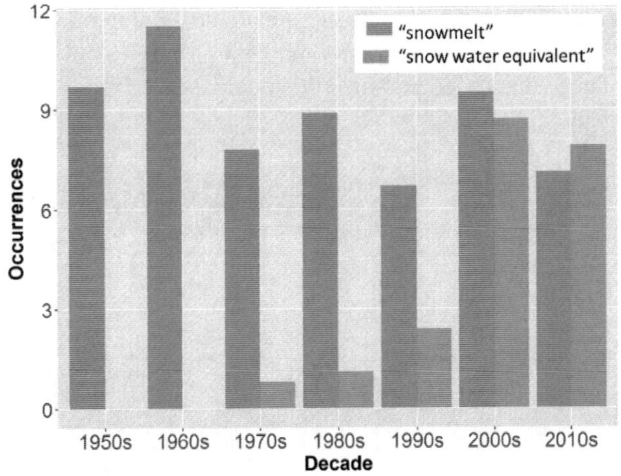

Figure 2. Variation of key term "snowmelt" and "snow water equivalent" through the history of ESC publications.

There is an overall trend is word usage that emphasizes a change in approach in understanding snowfall and snow on the ground. Early papers presented work on weather modification impacts, snowfall impacts on transportation infrastructure, and snow loading on buildings. More recently, papers have focused on process-based modeling and the variety of methods to determine large scale snow cover extent and water content. This demonstrates a fundamental change in the research presented at the Eastern Snow Conference, which had a heavy early focus on operational demands for snow data by engineers (e.g., Deslauriers, 1965), towards a more recent strong focus on process-based modeling and large-scale snow cover extent (e.g., Perry *et al.*, 2013).

Despite these changes in presentation focus, the Eastern Snow Conference remains a venue to present advances in technology for the observation and measurement of snow and ice, both with respect to *in situ* and remote sensing techniques. This results in these papers recording a detailed history of the evolution of how snow and ice are measured. This can be seen in papers detailing large, motor-driven frost measurement devices (Mason and Ruggles Jr., 1954) and the radioactive snow gauge (Gerdel, 1952) through to inexpensive and portable multiparameter snowpack sensors (De Roo *et al.*, 2017).

The most notable change through time in the direction of research in the Proceedings of the Eastern Snow Conference is the growth and dominance of remote sensing techniques. Of course, the early years of the ESC predate the widespread use of airborne and satellite-based sensors for snow measurement, though a rapid increase in papers using remote techniques was seen in the 1990s and continues through the present (Fig. 3). By the 2010s, "microwave" was the most commonly used word in paper titles. The increase in remote sensing-based research was accompanied by an increase in modeling-based research beginning in the 1980s (e.g., Chiang and Shen, 1982). Together, these research methods have overtaken *in situ* methods as the primary form of data collection. Although lamentable for some, this change is logical due to the increase in availability of satellite imagery, the increase in computing power, and the recognition of the high spatial variability of snowcover properties that are challenging to observe with *in situ* methods (e.g., Mann, 2018).

Lastly, spatial locator words in paper titles reveal an expansion from the tradition eastern North America home of the ESC to more globally diverse research (Fig. 4). Over the seven decades of publication of the Proceedings, changes to transportation infrastructure, global satellite coverage, and the growing online community have made it easier for researchers from all regions of the world to participate in research projects and conferences in any region. This has also resulted from a conscious expansion of the ESC to include Arctic researchers into the ESC as they did not have a forum of their own for research dissemination, and central North American research where the snowpack is more similar to that in the east than the mountain conditions to the west (B Goodison, *pers. comm.*).

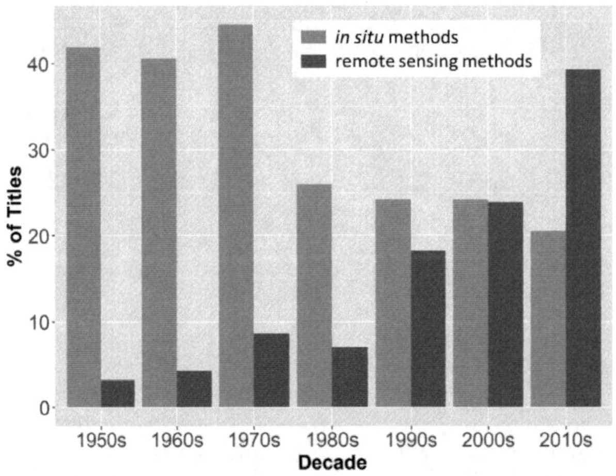

Figure 3. Variation in study methodology between *in situ vs* remote sensing research identified in ESC publications.

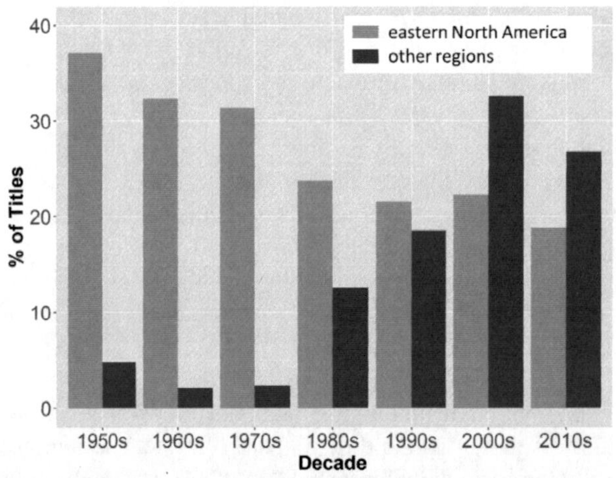

Figure 4. Variation in study spatial extent identified in ESC publications.

CONCLUSION

Since the 1950s, the Eastern Snow Conference has published papers and abstracts of the presentations at its annual meeting. These papers and abstracts represent a documentation of the history of snow and ice research and are used to highlight some of the changes observed in the discipline. These changes underline the growth and evolution of the snow and ice sciences and their response to concurrent changes in technology, scale, and, in some ways, funding availability. One of the most significant changes seen in this seven-decade record is the shift in participation from operations-based to academic-based researchers. While these two groups are often quite distinct, the research provided by each is complementary and important for the continual progression of putting theory into practice. At approximately the same time, though not necessarily correlated, was the increased use of satellite platforms for snow and ice research. Along with this shift was the expansion of the spatial scope of Eastern Snow Conference presentations to include pole-to-pole-to-pole research.

REFERENCES

Chiang LA, Shen HT. 1982. Numerical simulation of thermal-ice conditions in the upper St. Lawrence River. *Proceedings of the 40th Annual Eastern Snow Conference*, Toronto, ON: 24-36.

De Roo R, Haengel E, Rogacki S, Ekins C, Mousavi S. 2017. Inexpensive in-situ snow pack sensors for temperature, density, and grain size: a first look at the first season. *Proceedings of the 74th Annual Eastern Snow Conference*, Ottawa, ON: 35-45.

Deslauriers CE. 1965. Projects to alleviate ice-jams on the Chaudiere River in Quebec. *Proceedings of the 22nd Annual Eastern Snow Conference*, Hanover, NH: 12-24.

Feinerer I, Hornik K, Meyer D. 2008. Text mining infrastructure in R. *Journal of Statistical Software*, **25**(5): 1-54.

Fellows I. 2018. wordcloud: Word Clouds. R package version 2.6.

Gerdel RW. 1952. The development of the radioactive snow gage. *Proceedings of the 9th Annual Eastern Snow Conference*, Springfield, MA: 1-12.

Mann PC. 2018. *Spatial and temporal variability of the snow environment in the western Canadian Arctic*. MSc thesis, Wilfred Laurier University, Waterloo, ON, Canada. 95 pp.

Mason WC, Ruggles Jr FH. 1954. A very unique method of measuring frost depths in the ground. *Proceedings of the 10th and 11th Annual Eastern Snow Conferences*, Albany NY and Greenfield MA: 15-16.

Perry LB, Keighton SJ, Lee LG, Miller DK, Yuter SE, Konrad CE, Bryant MT. 2013. Synoptic influence on snowfall event characteristics in the southern Appalachian Mountains. *Proceedings of the 70th Annual Eastern Snow Conference*, Huntsville, ON: 141-157.

Microstructural Characterization of Mid-Latitude Snowpack through Micro-Computed Tomography

LAUREN B. FARNSWORTH[1] AND ZOE R. COURVILLE[1]

ABSTRACT

Understanding the physical and mechanical properties of snow is essential to determining the vulnerabilities of mid-latitude snowpack to climatic shifts. By improving our measurements of snow microstructure, we can refine our understanding of structural, thermal, and optical properties of snow. Micro-computed tomography is a non-destructive imaging technique that produces high-resolution images. We use a Bruker Skyscan 1173 micro-CT scanner housed in a -10 °C cold room to produce high-resolution (20 µm), two-dimensional, cross-sectional slices that are compiled and segmented to create three-dimensional visualizations of a snow. With these output files, we can calculate snow microstructural parameters such as: total porosity, mean grain size per sample volume, anisotropy, and surface-to-volume ratio with which we calculate specific surface area. These parameterizations can inform our understanding of the radiative properties of snow. These analyzed snow samples were collected in 2020 as a part of the SnowEx Intensive Observation Period in Grand Mesa, Colorado. In concert with the collection of these micro-CT samples, scores of snow measurements were taken, to which we will compare our microstructural parameterizations. Preliminary results show good agreement between ground penetrating radar, snow micropenetrometer, and optically-based microstructural data.

[1] Cold Regions Research and Engineering Laboratory, Hanover, NH, USA

Using Cosmic Ray Neutrons to Estimate Snow Water Equivalent in Prairie Environments

MADISON WOODLEY[1], ERIC SPROLES[2], AND SAMUEL TUTTLE[1]

ABSTRACT

Snow influences the water and energy fluxes for surface and groundwater systems in prairie environments. During the winter seasons, prairie soils are often frozen, with shallow, non-permanent snow cover. The expansive areas of low relief topography and little forest cover that are characteristic of prairie and agricultural areas allow wind to continuously redistribute and rework snow. This can lead to highly variable snow cover that ranges from deep snow drifts near barriers (e.g., trees, fences, breaks in topography) to bare, wind-scoured fields. Low vegetation such as crop stubble also has a strong influence on snow accumulation due to capture of wind-driven snow. This makes it challenging to accurately represent snow depth and snow water equivalent measurements. Additionally, snowpack can ablate rapidly in the prairie due to sublimation accelerated by high wind speeds and increased melting from weather fluctuations. For this reason, frequent snow observations are needed to track snowpack evolution and aid in prediction of flood events. We use a cosmic ray neutron sensor (CRNS) installed at Montana State University's Central Agricultural Research Center (CARC) in Moccasin, MT as part of SnowEx 2021 to assess the viability of CRNS for snow water equivalent (SWE) estimation in a prairie environment. CRNS measurements are sensitive to snow mass, which allows for continuous estimation of SWE. The large footprint of CRNS observations (approximately 200 - 300 m radius) allow it to integrate over a significant amount of snow spatial variability, which may enable SWE estimates that are more representative of the surrounding area than other continuous SWE observations (e.g., snow pillows). We compare CRNS-derived SWE estimates to manual snow measurements, continuous soil moisture observations, and drone-based LiDAR snow depth estimates within the sensor's footprint to assess the feasibility of CRNS for SWE monitoring in the prairie.

[1] Syracuse University, 900 South Crouse Ave, Syracuse, NY 13244, USA
[2] Department of Earth Sciences, Montana State University, Bozeman, MT 59717, USA

Impact of the Spatiotemporal Variability of the Snowpack Conditions on Liquid Water Fluxes

MICHEL BARAER[1]

ABSTRACT

The gap between the current knowledge of snowpack hydrological processes and the representation of those processes in hydrological models raises concerns about possible bias in long-term projections those models are used for. Improving those models requires a better quantification of the liquid water fluxes within the snowpack and a better qualification of the snowpack structure – flowthrough interrelations. This is especially true in non-mountainous environments where those aspects remain understudied.

Over the last winter, we have deployed a multi-method approach at the Ste Marthe experimental watershed in Quebec to study those questions. This approach combines photogrammetry, repetitive drone-based GPR surveys, multiparameter winter long monitoring as well as field measurements. So far, we have analyzed GPR results and combined those with 3D maps of the snow cover. Preliminary results are promising as we succeeded in isolating spatial variations in the hydrological reaction of the studied areas to rain-on-snow events. In addition, results so far show how combining 1.5 GHz drone-based GPR data to RBG of NIR image-based photogrammetry can produce bulk permittivity maps that provide repetitive insights for studying the hydrological behaviour of the snowpack. The next step will consist of comparing those permittivity maps to liquid water content measurements and tentatively produce series of snow moisture maps.

[1] Hydrology, Climate and Climate Change (HC^3) Laboratory, École de technologie supérieure, University of Quebec, Canada

Variability of Snow Depth Distributions in a Forested Mountain Basin from UAV-Lidar Remote Sensing

PHILLIP HARDER[1] AND JOHN W. POMEROY[1]

ABSTRACT

In complex terrain with heterogeneous vegetation, multiscale snow processes lead to large variability in snowpack distributions. Statistical representations of the snowpack are critical from many snowmelt modelling schemes but quantifying process-dependent variability across a complex basin in a comprehensive and consistent manner is challenging with traditional methods. The extent of snow surveys are limited by accessibility and labour limitations and forest cover frustrates most remote sensing approaches of snow depth. Lidar systems deployed on UAVs provide a means to resolve sub-canopy snow depth at centimeter accuracy, even in dense forests. This work seeks to demonstrate unique opportunities for UAV-lidar to quantify the variability in snow depth distributions at a research basin scale subjected to differences in accumulation, redistribution, interception, and ablation processes. To support this a sequence of UAV-lidar scans of Fortress Mountain Research Basin in the Canadian Rockies were taken in field campaigns in winter/spring 2019-2021. The results demonstrate differences in snow depth distribution as functions of multiscale controls on snow redistribution and ablation processes; these range from a dependency of tree wells to tree height during accumulation periods to differences in snow-vegetation interactions with slope and aspect during ablation periods. UAV-lidar observations provide unprecedented details on the spatial variability of snowpack's in complex terrain and forest cover as they accumulate and ablate that will be critical to developing understanding and improving snow modelling at all scales.

[1] Centre for Hydrology, University of Saskatchewan, Saskatoon, SK, Canada

Applying the Snow Characterization with Light and Temperature (SCLT) Method to better Understand the Evolution of a Winter Snowpack

ROSAMOND TUTTON[1] AND ROBERT WAY[1]

ABSTRACT

Snow cover duration and depth are critical metrics that are widely used for assessing global energy balance, ground thermal regimes, ecological modelling, and predicting cultural and infrastructure hazards in response to climate change. In Nunatsiavut, Nitassinan, and NunatuKavut (Labrador, Canada), expanding spatial and temporal snow depth measurement is imperative to understanding the future impacts of a changing climate. However, like many northern regions, logistics and the benefits of having a number of stations dispersed across diverse terrains limits the suitability of automated snow sensors such as SR50s. Low-cost methods for characterizing winter snow accumulation and ablation are sorely needed but commonly used methods including time-lapse imagery and iButton (temperature) snow stakes struggle with issues including weather conditions obscuring image acquisitions and the density-dependency of temperatures recorded in the snowpack. Recently, Tutton and Way (2021) proposed a new low-cost technique for monitoring snow thickness at remote field sites termed the Snow Characterization with Light and Temperature (SCLT) method. This approach uses vertically arranged light and temperature loggers and was initially deployed in 2019 at six remote field sites for evaluation. In subsequent years, additional SCLT stakes have been deployed across different ecotypes in Labrador and compared against time-lapse imagery. Here we report on follow up analysis investigating the utility of the SCLT method for characterizing snow depth including a discussion of suitable snow cover thresholds. These results will be used to better understand changes to snow in remote regions of Labrador and may facilitate cost-effective measurement of key snow variables in remote northern environments.

[1] Northern Environmental Geoscience Laboratory, Department of Geography and Planning, Queen's University, Kingston, ON, Canada

Global Determination of Snow Cover using Remote Sensing and a Near Real Time Processing Chain

SEBASTIAN RÖßLER[1] AND ANDREAS DIETZ[1]

ABSTRACT

Remote sensing offers the best prerequisites for obtaining comprehensive information on global snow cover. Although microwave remote sensing can provide limited information about the thickness of the snowpack and the water stored (i.e. snow water equivalent), its geometric resolution is not sufficient for an accurate spatial analysis. Optical remote sensing provides the required spatial resolution, but it is often compromised by clouds or at high geographical latitudes by the polar night. In order to obtain cloud-free information on daily snow cover from optical data, the German Aerospace Center developed the already established product Global SnowPack (GSP). It is based on the daily MODIS snow products originating from Terra and Aqua platforms and provided by NSIDC. With the help of sequential algorithms and additional data (digital elevation model, land cover classifications), pixels with clouds or polar night are continuously eliminated. While the Global SnowPack has so far only been calculated retrospectively for the entire hydrological year, there will now also be a near real time product (NRT-GSP). The latest MODIS data (these are available after approx. 2 days) are interpolated on a daily basis using the previous days. The product will be available in the future on the GeoService of the Earth Observation Center. We see an application of this product, for example, in the prediction of extreme hydrological events. In a recently published study, the development of the snow cover of various catchment areas of nival rivers derived from Global SnowPack was incorporated into a snowmelt runoff model. It was found that extreme high and low water events during the annual spring flood were reflected early in the development of the snow cover extent. With the help of the NRT-GSP product, such a development would be recognizable at an early stage and preparations could be made.

[1] German Aerospace Center, Oberpfaffenhofen, Germany

East *vs*. West: Contrasting Snowpack Properties in the Weddell Sea, Antarctica

STEFANIE ARNDT[1], MARCEL NICOLAUS[1], AND CHRISTIAN HAAS[1]

ABSTRACT

As Antarctic snow cover persists during most of the year, it contributes significantly to the sea ice mass and energy budgets due to comprehensive seasonal transition processes within the snowpack. The Weddell Sea has a particular role in this respect as it consists primarily of perennial sea ice in the west, while seasonal sea ice is dominant in the east, as reflected in the respective snow characteristics. However, the spatial and seasonal variability of the snow properties associated with, for example, prevailing atmospheric conditions and the corresponding influence on the underlying sea ice have not yet been investigated in detail.

To do so, we present here a comprehensive compilation of *in situ* snow pit observations of physical seasonal and perennial snowpack properties obtained during numerous expeditions in the Weddell Sea since the 1990s, covering spring, summer, autumn, and winter conditions.

First results show a highly heterogeneous snow microstructure in the eastern Weddell Sea in late summer, ranging from small wind slab crystals (<0.5 mm) to large depth hoar crystals (>4 mm). From north to south there is a dominant slush fraction in the north due to the relatively thin sea ice beneath. In contrast, observation further south indicate widespread temperature gradient snow metamorphism evidenced by a high fraction of depth hoar due to colder atmospheric conditions. In contrast, widespread wet snow metamorphism dominates the snowpack in the northwestern Weddell Sea at this time of year, indicated by a high fraction of melt-freeze clusters caused by latitudinally higher solar radiative fluxes. Preliminary seasonal comparisons in the northwestern Weddell Sea show higher mean bulk densities up to 350 kg m^{-3} associated with high proportions of compacted wind slab and faceted crystals in spring and early summer, while high proportions of depth hoar crystals result in lower bulk densities in winter and early autumn.

Combining these snow observations with atmospheric and oceanic conditions along the previous drift trajectories of the sampled ice floes will provide substantial insight into the associated regional difference in the seasonal sea ice mass and energy budgets in the Weddell Sea.

[1] Alfred-Wegener-Institut Helmholtz-Zentrum für Polar-und Meeresforschung, 27570 Bremerhaven, Germany

High Resolution Spatial Estimates of Average Snow Density and Snow Water Equivalent from Differenced LiDAR Elevations and GPR Travel-Times at Grand Mesa, Colorado

TATE G. MEEHAN[1,2], AHMAD HOJATIMALEK[1], H.P. MARSHALL[1], ELIAS J. DEEB[1,2], DAN MCGRATH[3], RYAN WEBB[4], AND RANDALL BONNELL[3]

ABSTRACT

Pit, ground based radar, and airborne LiDAR observations of the snowpack were acquired throughout the SnowEx 2020 intensive observation period at Grand Mesa, Colorado. We developed a method for automatically determining the ground-penetrating radar (GPR) wave two-way travel-time for the reflection off of the ground through the snow cover. Our method uses the maximization of the coherence between co-polarized and cross-polarized GPR channels. We validated the accuracy of these travel-time picks by comparing the snow depth measured by differencing the snow-on and snow-free airborne LiDAR acquisitions to the snow depth estimated from the GPR travel-times and the average density of nearby snow pits which were converted to radar wave speed ($R = 0.67$, RMSE = 12 cm). To extend this result, using multivariate linear regression, we developed a model that, once trained using the LiDAR and GPR derived snow density, predicts the average snow density within the domain of the LiDAR acquisition without dependence on GPR travel-times. The modeled density shows intriguing spatial variability that agrees well with the spatial observations of average snow density from the LiDAR-GPR method ($R = 0.75$, RMSE = 22 kg/m^3). Using the modeled densities and the LiDAR measured snow depths, we estimated the snow water equivalent (SWE), and found good agreement with the SWE measured at 96 snow pits within the LiDAR domain ($R = 0.73$, RMSE = 39 mm).

[1] Department of Geosciences, Boise State University, Boise, ID, USA
[2] U.S. Army Cold Regions Research and Engineering Laboratory, Hanover, NH, USA
[3] Department of Geosciences, Colorado State University, Fort Collins, CO, USA
[4] Department of Civil, Construction, and Environmental Engineering, University of New Mexico, Albuquerque, NM, USA

Streamflow Generation and the Importance of Atmospheric Rivers to Annual Flooding for the Coupled Wolverine Glacier-Creek System, Kenai Mountain, Alaska

TODD GROTE[1], ALEX CRAWFORD[2], AND AARON JACOBS[3]

ABSTRACT

Streamflow generation in glacierized basins may be produced by physical processes including temperature-induced melt, rainfall, rain-on-snow (ROS), glacial lake outbursts, or ice jams. The timing of annual maximum flows (floods) is dependent upon a combination of ablation season meteorological, hydrological, and glaciological conditions. Here we investigate climatological aspects of the hydrology of the coupled Wolverine Glacier-Creek system located in the humid, maritime climate of the Kenai Peninsula, Alaska. Hydrological, climatological, and glaciological datasets for Wolverine Glacier and Creek were used to investigate proglacial streamflow generation, and the timing and causes of annual flooding.

Wolverine Creek's hydrologic system is a classical nival system distinguished by three distinct phases of streamflow generation: snowmelt-dominant; ice-dominant; and rainfall-dominant. Although much of the proglacial streamflow is generated by snow and ice melt, most annual floods are associated with rainfall coupled with melt, commonly in September (~45%) and August (~29%). Many annual floods occur well after the transient snow line was at or very near the long-term ELA, and after the peak contribution of ice melt-dominated streamflow. Daily and hourly re-analysis output show advection of warm, humid air into the Kenai Mountains around a center of low pressure generally situated over, or just off the coast of, south-central Alaska is linked to annual flood generation regardless of timing. Moisture advection from the northern and sub-tropical Pacific Ocean by strong low-level (850 mb) winds manifests as atmospheric rivers (ARs) of anomalously high precipitable water, which, along with orographic enhancement, provide ample moisture to trigger annual floods at Wolverine Creek and throughout the region. A review of the fragmented annual flood series for Wolverine Creek indicates that since 1980, 19 of 20 (95%) of annual floods were triggered by ARs and suggests that they need to be considered in glaciological and hydrologic studies.

[1] Geoscience Program, School of Natural Sciences, Indiana University Southeast, New Albany, IN, USA
[2] Centre for Earth Observation Science, Department of Environment and Geography, University of Manitoba, Winnipeg, MB, Canada
[3] National Weather Service, Juneau, AK, USA

Snow Measurements from the First Two Years of the Coastal Labrador Climate and Weather Monitoring Program

YIFENG WANG[1], ROBERT WAY[1], ROSAMOND TUTTON[1], AND JORDAN BEER[1]

ABSTRACT

Snow indices are key inputs for numerical weather prediction and are critical for understanding the impacts of climate change on Subarctic and Arctic environments. In Canada, recent government-led changes to snow monitoring infrastructure have reduced the availability of long-term snow cover records. Diminished snow monitoring in northern Canada can be particularly challenging for residents of remote Indigenous communities who rely heavily on suitable snow conditions for safe winter travel. The Labrador region of northeastern Canada (294,330 km^2) has been disproportionately impacted by recent changes to snow monitoring, having lost snow measurement equipment at three (Cartwright, Churchill Falls, Wabush) out of the five stations that were previously recording snow depths in 2010. In summer 2019, the Coastal Labrador Climate and Weather Monitoring Program (CLCMWP) was launched through a partnership between Queen's University, the Labrador Institute, the Government of Newfoundland and Labrador, the Nunatsiavut Government, the NunatuKavut Community Council, and the Mushuau Innu First Nation in response to community needs for improved climate and meteorological monitoring in coastal Labrador. CLCMWP's main contribution has been the installation of six remote automated weather stations (RAWS) in the communities of Postville, Rigolet, North West River, and Black Tickle-Domino, and at two locations along the Trans Labrador Highway (near Cartwright Junction and the community of Red Bay). Each RAWS is equipped with sensors for air temperature, barometric pressure, wind speed, wind direction, relative humidity, precipitation, snow depth, soil moisture, soil conductivity, soil temperature, and solar irradiance. Monitoring over two full snow seasons (2019-2020 & 2020-2021) and disparities with currently available snow monitoring products highlight the importance of having dense networks of RAWS in northern coastal regions. Finally, we offer some brief perspectives on the cryospheric and community impacts of the anomalous weather conditions of winter 2020-2021 as observed at CLCMWP stations throughout Labrador.

[1] Northern Environmental Geoscience Laboratory, Department of Geography and Planning, Queen's University, Kingston, ON, Canada

SNO-FOO 2021 – Not awarded

Past Recipients

The Eastern Snow Conference annually bestows upon a distinguished snow scientist who, in striving for excellence in snow research, contributes to an event of notable humor the highly coveted Sno-Foo Award.

Year	Winner	Affiliation
2021	*Not awarded*	
2020	*Meeting cancelled*	
2019	Alexandre Roy	Université du Québec à Trois-Rivières
2018	Alexandre Langlois	Université de Sherbrooke
2017	Michael Durand	The Ohio State University
2016	Roger De Roo	University of Michigan
2015	Kevin Côté	Université de Sherbrooke
2014	Dorothy Hall	NASA GSFC
2013	Benoit Montpetit	Université de Sherbrooke
2012	Don Pierson	NYC DEP
2011	Ken Rancourt	Mt. Washington Observatory
2010	Kyung-Kuk (Kevin) Kang	University of Waterloo
2009	Rob Hellström	Bridgewater State University
2008	Steven Fassnacht	Colorado State University
2007	The Group of 9[1]	
2006	Andrew Klein	Texas A&M University
2005	Claude Duguay	University of Alaska-Fairbanks
2004	Chris Derksen	Meteorological Service of Canada
2003	Miles Ecclestone	Trent University
2002	Danny Marks	U.S.D.A.
2001	Brenda Toth	University of Saskatchewan
2000	Mauri Pelto	Nichols College
1999	Ross Brown	Meteorological Service of Canada
1998	Mary Albert	CRREL
1997	Jean Stein	Université de Laval
1996	Colin Taylor	Trent University
1995	Mike Demuth	NHRI
1994	Bert Davis	CRREL
1993	John Pomeroy	NHRI
1992	Tom Niziol	National Weather Service
1991	Terry Prowse	NHRI
1990	Kersi Davar	University of New Brunswick
1989	Gerry Jones	INRS-EAU
1988	Robert Sykes	SUNY
1987	John Metcalfe	Meteorological Service of Canada
1986	Peter Adams	Trent University
1985	Don Wiesnet	National Weather Service
1984	Barry Goodison	Meteorological Service of Canada

[1] Jimmy MacDonald (University of Saskatchewan), Bill Floyd (University of British Columbia), Chris DeBeer (University of Saskatchewan), Wendell Koning (Government of Alberta), Jaime Hood (University of Calgary), Danika Muir (University of Calgary), John Jackson (University of Calgary), Sarah Forte (University of Calgary), Masaki Hayashi (University of Calgary)

LIST OF ATTENDEES

Mark Adams	*Austrian Research Centre for Forests*
Claudio Antoni	*University Ca'Forscari of Venice*
Anselm Arndt	*Humboldt-Universitat ze Berlin*
Stefanie Arndt	*Alfred Wegener Institute*
Gifty Attiah	*Wilfred Laurier University*
Michel Baraer	*University of Quebec*
Carsten Becker	*University of Innsbruck*
Jordan Beer	*Queen's University*
Andre Bertoncini	*University of Saskatchewan*
Alice Bradley	*Williams College*
Mary Jo Brodzik	*University of Colorado*
Dominique Brunet	*Environment and Climate Change Canada*
Ella Bump	*Colorado State University*
Yueqian Cao	*Duke University*
Alex Cebulski	*Vancouver Island University*
Eunsang Cho	*University of New Hampshire*
Krystopher Chutko	*University of Saskatchewan*
Harry Clayton	*University of Durham*
Riley Cormier	*Carleton University*
Zoe Courville	*Cold Regions Research and Engineering Laboratory*
Kenneth Curry	*National Oceanic and Atmospheric Administration*
Sophia D'Aurora	*Environment and Climate Change Canada*
Elias Deeb	*Cold Regions Research and Engineering Laboratory*
Vasana Dharmadasa	*Université du Québec à Trois-Rivières*
Jonathan Edwards-Opperman	*U.S. National Ice Center*
Richard Essery	*University of Cambridge*
Lauren Farnsworth	*Cold Regions Research and Engineering Laboratory*
Abbas Fayad	*University of Saskatchewan*
Anika Forget	*Queen's University*
Bart Forman	*University of Maryland*
Richard Forster	*University of Utah*
Abror Gafurov	*German Research Centre for Geosciences*
Camille Garnaud	*Environment and Climate Change Canada*
Hayden Gillooly	*Williams College*
Barry Goodison	*Environment and Climate Change Canada (Retired)*
Todd Grote	*Indiana University Southeast*
Grant Gunn	*Michigan State University*
Dorothy Hall	*ESSIC/University of Maryland*
Phillip Harder	*University of Saskatchewan*
Sean Helfrich	*National Oceanic and Atmospheric Administration*
Warren Helgason	*University of Saskatchewan*
Jennifer Jacobs	*University of New Hampshire*

Vicki Jagdeo	*Lehigh University*
Kamil Kachniarz	*University of Silesia*
DK Kang	*NASA Goddard Space Flight Center*
Richard Kelly	*University of Waterloo*
Hadi Mohammadzadeh Khani	*Université du Québec à Trois-Rivières*
Edward Kim	*NASA Goddard Space Flight Center*
Rhae Sung Kim	*NASA Goddard Space Flight Center*
Joni Kincaid	*Texas A&M University*
Fraser King	*University of Waterloo*
Joshua King	*Environment and Climate Change Canada*
Christophe Kinnard	*Université du Québec à Trois-Rivières*
Andrew Klein	*Texas A&M University*
John Kochendorfer	*National Oceanic and Atmospheric Administration*
Jonas Kohler	*German Aerospace Center*
Daniela Krampe	*Alfred Wegener Institute*
Theodore Letcher	*Cold Regions Research and Engineering Laboratory*
Jewell Lund	*Cold Regions Research and Engineering Laboratory*
John MacPhee	*Meteorological Service of Canada*
Sarah Marcil	*Environment and Climate Change Canada*
Chris Marsh	*University of Saskatchewan*
HP Marshall	*Boise State University*
Mariah Matias	*Lehigh University*
Alex Mavrovic	*Université de Sherbrooke*
Rachel McCrary	*National Center for Atmospheric Research*
Colin McLaughlin	*University of Maryland*
Tate Meehan	*Boise State University*
Pragya Mehrishi	*Czech University of Life Sciences*
Eva Mekis	*Environment and Climate Change Canada*
Colin Meyer	*Dartmouth College*
Dori Miller	*University of Saskatchewan*
Justin Minder	*University of Albany*
Alireza Moghaddasi	*University of Maryland*
Jhon Mojica Moncada	*NYU Abu Dhabi*
Benoit Montpetit	*Environment and Climate Change Canada*
Colleen Mortimer	*Environment and Climate Change Canada*
Pat Naple	*University of Albany*
Bibhash Nath	*Hunter College*
Wenge Ni-Meister	*City University of New York*
Carrie Olheiser	*National Weather Service / National Center for Atmospheric Research*
Jason Olivier	*Cold Regions Research and Engineering Laboratory*
Julie Parno	*Cold Regions Research and Engineering Laboratory*
Mauri Pelto	*Nichols College*
John Pomeroy	*University of Saskatchewan*
Alicia Pouw	*Wilfred Laurier University*
Hamish Pritchard	*British Antarctic Survey*

Holly Proulx	*University of New Hampshire*
Mark Raleigh	*Oregon State University*
Joan Ramage	*Lehigh University*
Kenneth Rancourt	*Mount Washington Observatory*
George Riggs	*Science Systems and Applications, Inc.*
David Robinson	*Rutgers University*
Alvaro Robledano	*Grenoble Alpes University*
Sebastian Roessler	*German Aerospace Center*
Alexandre Roy	*Université du Québec à Trois-Rivières*
Benedita Santos	*Institute for Earth Observation*
Forrest Schoessow	*Ohio State University*
Maria Silver	*Broomfield High School*
Siddharth Singh	*Duke University*
Paul Siqueira	*University of Massachusetts*
Craig Smith	*Environment and Climate Change Canada*
Ulrich Strasser	*University of Innsbruck*
Ahmet Emre Tekeli	*Igdir University*
Ezequiel Toum	*Instituto Argentino de Nivologia*
John Toupin	*University of Alberta*
Samuel Tuttle	*Mount Holyoke College*
Rosamond Tutton	*Queen's University*
Vincent Vionnet	*Environment and Climate Change Canada*
Carrie Vuyovich	*NASA Goddard Space Flight Center*
Branden Walker	*Wilfred Laurier University*
Lizhao Wang	*University of Maryland*
Wifeng Wang	*Queen's University*
Michael Warscher	*University of Innsbruck*
Madison Woodley	*Syracuse University*
Melissa Wrzesien	*NASA Goddard Space Flight Center*
Huaxia Yao	*Ministry of Environment and Climate Change*
Puneeth Yogananda	*University of Michigan*
Qingping Yu	*University of Saskatchewan*
Wenyu Zhao	*Lanzhou University*
Jiyue Zhu	*University of Michigan*